The Price of Water

And the same night there fell a shower of rain,
For which their mouths gaped like the cracks of earth
When dried to summer dust. Till taught by pain,
Men really know not what good water's worth.
If you had been in Turkey or in Spain,
Or with a famished boat's crew had your berth,
Or in the desert heard the camel's bell,
You'd wish yourself where truth is in a well.

Lord Byron, *Don Juan*

The Price of Water

Studies in Water Resource Economics and Management

Stephen Merrett

IWA Publishing

Published by IWA Publishing, Alliance House, 12 Caxton Street, London SW1H 0QS, UK

Telephone: +44 (0) 20 7654 5500; Fax: +44 (0) 20 7654 5555; Email: publications@iwap.co.uk
Web: **www.iwapublishing.com**

First published 2005

Typeset by Gray Publishing, Tunbridge Wells
Printed by TJ International (Ltd), Padstow, Cornwall

British Library Cataloguing in Publication Data
A CIP catalogue record for this book is available from the British Library

Library of Congress Cataloging- in-Publication Data
A catalog record for this book is available from the Library of Congress

ISBN 1 84339 081 7

Contents

Preface

On the evening of 22 September 1994, I received a telephone call from a Danish management consultancy asking me if I would be free to work in Latvia for a month on the economics of a local water utility. My reply was positive, but I was compelled to add that I knew absolutely nothing about water or water utilities. The consultant replied that a couple of months would elapse prior to the project's start-date; from previous contact he was sure that if I applied myself to the subject-area during those 8 weeks, all would be well. The deal was struck. I raised my eyes to the heavens (well, the ceiling of my flat) and said: 'Let it be water'.

During the previous 30 years, my entire working life had been spent carrying out economic research in a variety of applied fields. These included: higher education and student finance, the production of fertilizers, the British civil space programme, international trade and capital flows, housing construction and finance, and land-use planning.

Taking early retirement from University College London in 1994, I decided to re-invent myself as an environmental consultant. My two daughters' persuasive arts, and my lifelong love of the natural world, were the roots of this choice. But environmental research embraces an extraordinary variety of subjects; so it was that chance telephone call from Denmark that made water resources my focus. By February 1995 I was on the River Daugava in Daugavpils, Latvia's second city, carrying out an affordability study of price increases for the town's water and waste water services. Moreover, by this stage, I had already spent 5 months preparing a new book on the economics of water resources, which was later published in 1997.

In my view, the oldest and still the most important production activities of humankind are securing food and water, building and maintaining housing, making clothing, and providing health care and education. These activities can be thought of as forms of *reproductive production*. Each is simultaneously a form of human labour as well as a direct and necessary condition for the reproduction of the species. The fact that more than half my working life had passed in the study of the economics of education and the economics of housing made the shift into the study of water resources that much easier.

1

Introduction: getting to grips with water

1.1 INSTREAM AND OUTSTREAM

This volume brings together 14 papers, on the economics and management of water resources, which were first published in various professional journals between 1999 and 2004. The subject matter of water resources research can be divided into two (interdependent) blocks. The first block concerns instream water, that is, the flows of springs, streams and rivers, the lakes these flows create, and the immense hidden stores of groundwater that lie beneath our feet as aquifers. This is the world of the hydrological cycle. The second block is composed of the flows of water abstracted from instream sources, stored, distributed to water users and then discharged back to the instream medium. This is the world of the hydrosocial cycle (Merrett 1997: 6–7).

The principal research areas in the case of instream water are hydroelectric power, navigation, fishing, conservation, recreation and environmental tourism. In contrast, for outstream water, research focuses on the capture of instream sources, their consequent channelling, storage, treatment of raw water, use by households and farmers and urban institutions, sewerage, the treatment of waste water and its discharge back to the instream medium or to coastal waters. This volume's collection of papers mainly addresses outstream themes.

The reader will soon be aware that I have a strong interest in the philosophy of science. Research programmes, in my view, should combine theory, method, fieldwork and policy development in a continuous interplay. To take four examples, Chapter 13, on the domestic demand for water services in Africa, is a paper in method. Chapter 9 on the urban market for farmers' water-rights is an exercise in theory. Chapter 4 on farm-level drought management in England's Anglian Region has a major field-work content.

Chapter 15 on the interaction between economic instruments and environmental regulation with respect to waste water discharges, addresses policy development.

1.2 REUSE AND RECYCLING

The fundamental distinction between instream and outstream water is complemented by the related distinction between the reuse of water and the recycling of water. Reuse and recycling are two concepts repeatedly drawn on in this volume as well as in my two books on water resources economics and management (Merrett 1997, 2002a).

Reused water is waste water and irrigation drainage that, *prior to its return to the instream resource*, is captured and used again (perhaps repeatedly). Reuse can take place *within* a single institution (such as a sugar-mill) or it can occur *between* institutions as when urban waste water is reused for irrigation purposes. Reuse may have real advantages, such as lowering abstraction costs or by reducing waste water treatment costs.

Recycling refers to water that is abstracted, used by households, industry, etc., and then the fraction that is not consumed as evapotranspiration flows back to the catchment's rivers, aquifers and lakes. The importance of recycling to hydroeconomists is that it augments the hydrological resource from the point at which the recycling occurs. The negative characteristic is that recycled flows may pollute the resource. The proportion of water used that is consumed, and therefore is unavailable for recycling, varies between categories of use.

1.3 CATCHMENT AND REGION

Water resources research is carried out either at the catchment scale or at the regional scale, where 'region' refers to any area with a defined boundary. The beauty of working at the catchment (river basin) scale is seen to be the unity of its hydrological flows – precipitation, evapotranspiration, run-off and groundwater recharge. However, groundwater boundaries may not fall within the catchment boundary. Moreover, hydrosocial flows such as the import of water from another catchment or the export of waste water beyond the catchment also undermine the supposed unity of flows.

The attraction of regional analysis is its flexibility. It can be applied to a single dwelling, a village, a city, a province, a country, an island – or what you will. However, a region will usually be part of a river basin, or overlap two basins, or contain several basins, so that the linkage of hydrological and hydrosocial flows becomes extremely complex. The hydrological unity of the single river basin is lost.

In 1996 I began developing a method concept now referred to as 'the hydrosocial balance', which is applicable at the catchment or at the regional scale. It is a planning tool that appears in many of this volume's papers and, in its most complete form, in Chapter 3, Table 3.1. My own fieldwork in applying the concept has been carried out for a single house in southern Spain, the island of Jersey, Gaza, the West Bank of Palestine, the Thames catchment in England and a farm in the Anglian Region. As is argued in Chapter 3, the hydrosocial balance offers excellent potential for the regional-scale planning of water's civil engineering infrastructures and their capital financing.

1.4 CATCHMENT DEFICITS

It is likely that the paper destined to have the greatest professional impact on theory, fieldwork and policy development is Chapter 8, 'Catchment water deficits in the twenty-first century'. This is my favourite child. Its objective is to provide a general theory of how the water resources of a river basin shift from surplus to deficit and the means by which water resource institutions can manage or reverse this shift.

The article bites the bullet of defining surplus and deficit in the following way. A catchment water surplus is a situation in which, throughout the course of a specified year, total precipitation in the basin is sufficient to simultaneously satisfy four conditions:

1. Abstraction from the aquifer is maintained at a sustainable rate.
2. Outstream water fully meets the economic demand for water from households, agriculture, mining, manufacturing, construction and the services sectors.
3. The basin population's economic demand for food is fully met from domestic rainfed and irrigated farming or from domestic fisheries or from food imports financed by the basin's commodity and service exports.
4. The river's instream flows do not fall below defined minima.

When a catchment water surplus does not exist, the river basin is in deficit.

The analysis in Chapter 8 proceeds by illustrating the argument with a fictitious catchment called the Dwyer, which in the base-year is in surplus and in which total population increases by one-third every 25 years, that is, about 1.15% per annum. Moreover, output per capita is also rising. This combination of growth both in population and economic productivity, associated with the rise of world capitalism, is the source of deep unease amongst the professional staff of the Dwyer Catchment Authority. The Authority estimates the catchment's economic output in 2025. After assuming that the ratio of outstream water use to basin output is constant, the Authority thereby derives an estimate of total water use in 2025. The Authority anticipates that, for the first time in its history, in 2025 the catchment will move into a water deficit; one or more of the four necessary conditions for surplus will have been breached.

The paper then reviews 12 redemptive options (see Chapter 8). Six of these moderate the situation but do not prevent entry into water deficit; examples are importing water from another catchment, and the extension of water reuse. Six other options *do* prevent entry into water deficit; examples are reducing the rate of growth of population, and increasing water's productivity in terms of value added per cubic metre of water consumed.

Chapter 8 re-states the implications for a river basin of finding itself in water deficit. These are one or more of the following situations:

1. The basin is pumping its groundwater at an unsustainable rate.
2. There is insufficient outstream water to meet the economic demand for it.
3. The population has to import water, or food that it is unable to pay for from its exports of goods and services.
4. The basin's citizens must accept the economic and environmental losses following from its river diminishing in volume.

The nightmare scenario is a river basin in which groundwater is being exhausted, households, farmers and other actors cannot purchase the water they require, food imports cannot be paid for, making the basin dependent on powerful allies, and the river has been destroyed.

The article ends with suggestions for catchment research projects that establish whether or not any single river basin is in deficit and, if so, what are the causes and the policy options.

1.5 THE VIRTUAL WATER CONTROVERSY

In the previous section, one of the criteria for surplus at the river basin scale is that the basin population's economic demand for food is fully met from domestic rainfed and irrigated farming, or from domestic fisheries, or from food imports financed by the basin's commodity and service exports. The nightmare scenario for a deficit catchment includes a situation where food imports cannot be paid for, making the basin dependent on powerful allies; the relation between Egypt and the USA is an example. In fact, it was Tony Allan's work on the hydropolitics of the Middle East that first led me to include the water needs of domestic food production in the definition of water surplus (Allan 2001).

During the 1990s (and even prior to that decade) Allan had developed a concept that he now refers to as 'virtual water'. Virtual water is defined as the water needed to produce agricultural commodities, particularly in a context where they are exported. The concept is now widely used in the discussion of the need for water in food production, particularly in the arid, low-income nations.

Chapters 6 and 7 develop a strong critique of the virtual water concept, making the case that it be abandoned. The core of Allan's argument is that when a Region A discovers that the crop water requirements of food self-sufficiency are impossible to satisfy, the consequent water deficit can be resolved by the import of virtual water from Region B. To indicate the scale of these virtual water imports, we can take the example of wheat. One tonne of exported wheat requires about 1000 tonnes of virtual water (Allan 2001: 106, 126). Less than 0.1% of the virtual water is physically embedded in the food grains themselves. During cultivation in Region B, more than 99.9% of the virtual water returns to the irrigation cycle as farmland drainage or is lost in evapotranspiration. As Allan writes (2001: 106): 'At the 1000 tonnes (cubic metres) of water per tonne of grain estimate of water content the [MENA] regional imports of virtual water by the mid-1980s were equivalent to the annual flow of the Nile into the Egyptian agricultural sector.'

My critique is four-fold.

First, the term is redundant; virtual water is nothing more or less than the water needed to produce agricultural commodities.

Secondly, there is absolutely nothing virtual about virtual water.

Thirdly, when one approaches agriculture from the perspective of water resources there is a danger that the experience of farming is seen largely with respect to its crop water requirements. As a result, a more rounded vision is lacking, one that understands that the water theme is only one amongst many, such as soil characteristics, land rights, labour skills, pest control, farm budgets and product markets. Consequently, if we use the term 'the import of food', this opens up major questions rendered invisible by 'the import of virtual water'. Have food imports led to higher population birth rates in water deficit regions than would have occurred in their absence? Do food imports weaken the farm sector of the importing country? Do food imports open the importing country to political control from the exporting country? Will the importing country be able to maintain its foreign exchange expenditure on food imports in the long run? Finally, if food imports are subsidized when they are shipped from the European Union and North America, will these subsidies be maintained in the long term?

Fourthly, the confusion of water with food that accompanies the term 'virtual water' fosters analytic errors such as that food grain purchases are transported water, that food imports remedy water deficits in the importing countries and that imported food brings 'total water self-sufficiency' for the importing region (Allan 2003: Figure 3). *The greatest weakness of Allan's concept is that it repeatedly confuses crop outputs with the water required to produce them.* If, as Allan suggests, Hoekstra and Hung (2002) conclude that 695 km^3 of virtual water is traded each year, why have none of us seen the boats within which it is carried? Why are they not causing gridlock in the world's sea channels? Virtual water is real water; if it is indeed traded it needs to be transported to the new owner. In fact, no such trade ever takes place in the case of food exports. Hoekstra and Hung, outstripping Allan, commit a 695 km^3 error.

In fact, the huge economic, political and social processes that are addressed by Allan in terms of 'virtual water' can be reset in a world where arid and semi-arid regions (and others) do not have the capacity to feed their populations and so import food. These imports mean that less food production and therefore less water is required in these regions' irrigated agriculture. Where the region's exports are insufficient to finance food imports, regional politicians may deflect attention from such dependence; the availability of imported food allows them to postpone new water supply initiatives, to delay difficult decisions about the demand management of their water resources and to neglect the issues of birth control.

1.6 FOUR INTERPRETATIONS OF DEMAND

Among Chapter 8's redemptive options for catchment water deficits, the tenth concerns the effective demand for water. During the years that I worked on Palestinian water resource challenges, one of the draft outputs – co-written with my colleague Eng. Khalil Saleh – was a review of a number of documents assessing the future water demand of the Palestinians. In the course of this work it became clear that the term 'the demand for water' is used with a confusing variety of meanings. For an economist with an unslaked thirst for the philosophy of science, the subject seemed to be worth pursuing for the most practical of reasons. The outcome was a short paper entitled 'The demand for water: four interpretations', published as Chapter 10.

The core of the argument is that 'the demand for water' may refer to:

1. the use of water;
2. the consumption of water, that is, the evaporative losses during use;
3. the need for water; and
4. the economic demand for water, represented by the economist's trusty 'demand function' that relates quantity purchased and unit price.

This confusion of meanings, one with another, has consequences. To take demand to mean both *use* and *consumption* neglects the truth that a given volume used can be associated with wide variations in the volume consumed. This difference is at the core of much current debate in the field of irrigation engineering as the discussion lists show. To take demand to mean both *use* and *need* obscures the shortfall of use against need for more than half the world's population. To take demand to mean both *use* and the *economic demand* for water confuses the use of water with one of its determinants, whenever costs borne by the user vary with the volume used.

In the hydrosocial balance, the categories of supply include the leakage and evaporation losses that occur between the point at which human society appropriates outstream water and the point of delivery of water to user properties. Curiously such losses are almost always treated in the current literature as a form of 'demand'. The result is that consultants' forecasts of growth in 'demand' include (as unaccounted-for-water – UFW) supply-side leakage and evaporation. Furthermore the reduction of these losses in abstraction, storage and distribution prior to the delivery of water to the user is said to be a form of 'demand management'. This does not seem helpful. A manufacturer of refined sugar, when considering losses from output because of pilfering, or contamination while in the warehouse or destruction in a road or rail accident en route to the supermarket, would never regard this as a demand for sugar, a bizarre act of use by a consumer whom the sugar never reaches. The manufacturer would regard all of these as storage or distribution losses in the supply chain. So should it be with the supply of water.

There is a more general point here. Water resource management is now widely seen as principally a form of 'demand management'. To me such an approach seems nonsensical. This is especially the case when so much 'demand-management', as with reuse and the reduction of losses, turns out to be made up of supply-side initiatives. With outstream water we should always integrate in our thinking and our practice both demand-side *and* supply-side strategies.

1.7 PRICING THE RESOURCE

When a household, firm or farmer wishes to access water, the costs incurred by the user can take a variety of forms. The first form, the joy and delight of the economist, is that of a price paid to the supplier (such as a water utility) per unit quantity. For policy-makers committed to demand management in England, for example, a most encouraging development in the last two decades is the increase in the proportion of households that are on a priced, metered supply – from about 2% of the total number of households to about 20%.

The second form again directly relates volume received to cost incurred by the user. But in this case it is because the user is also the supplier of the water. This can be called the own-supply case. Karin Kemper, for example, describes the situation in the Curu catchment of Brazil where farmers use electric motors to pump irrigation water from the river. She describes the importance of electricity tariffs to irrigators' costs (Kemper 1996).

The third form in which users incur costs for access to water is where there is only an *indirect* link to volume used. For example, in Archangel in northern Russia where I was working in 2004, household payments for water used are based on a fixed tariff in roubles/m^3 and on the assumption that households use 300 litres per capita per day (lcd). This charge per person is then multiplied by the number of family members registered as living at a given address. In fact, *no one knows what is the volume of use per person*. The 300 lcd may be wildly inaccurate. However, the variation of the tariff paid with the number of family members clearly does have an indirect relation to volume used.

The fourth form of cost-use relation is where the payment made is a fixed charge for the user, invariant with volume. In England's districts, in cases where households are not metered, they pay a water and waste water charge based on the value of the property in which they live.

In this book a number of the papers deal with charging for water. Chapter 4 has a case-study of the price of water charged by an English regional water utility to a local farmer and the alternative costs of water incurred were he to begin abstraction from a local drainage channel – the own-supply case. Chapter 5 is a market-clearing proposal for raising water prices and fixed charges in the management of regional drought. The *political* feasibility of introducing such a management tool turns on (i) the proportion of households that have metered use and (ii) the protection of low-income families. This is the first time that water pricing has been suggested for drought management alongside the familiar informational, infrastructural and regulatory instruments. Chapter 11 reviews alternative methods of designing abstraction charges. Chapter 12 sets out 12 theses on the interrelation of the cost of irrigation water to farmers and the volume of water used. Chapter 15 discusses the design of waste water charges for sewage collection, treatment and disposal.

1.8 STUDIES OF HOUSEHOLD WATER USE AND THE WILLINGNESS-TO-PAY FOR WATER

A fascinating area of research in low-income countries in the past 20 years, indelibly associated with the names of Dale Whittington and his colleagues, is that of detailed case-studies of families' water use behaviour and their observed ability and willingness-to-pay for water. The first strength of the water demand school is its demonstration of how socially complex the networks of access and distribution can be. The second strength is the detailed examination of the dynamic and competitive markets in water that exist in many of Africa's villages and towns. The third strength of the water demand school is its success in estimating the proportion of domestic income absorbed by water purchases. At the same time, there are major weaknesses of the water demand school's work. The principal one is that the authors take a naturalistic standpoint in which the research into these communities' behaviour is limited to the actions of silent men, women and children, moving across a landscape like so many ants in a natural-historical study of a savannah colony. Household members are not treated as intelligent, resourceful, purposive and reflexive citizens. These arguments are made in Chapter 13.

Chapter 14 critically reviews the deployment of the willingness-to-pay concept in 11 studies published by Whittington and his fellow workers in the years 1988–98. The countries visited were Haiti, Ghana, Nigeria, Pakistan, the Philippines and Uganda. The critique is drawn from the branch of philosophy known as semiotics and it ends with an alternative approach to field research into households' economic demand for water. The paper includes the contrasts I drew between the 'official version' and the 'true story' of the price of water paid by a family in Yerevan, Armenia; this is a norm-based system like the one in Archangel already referred to.

1.9 TEARING UP WATER AND FLOATING ON WATER-RIGHTS?

Water is a collection of molecules, each of which consists of two atoms of hydrogen bonded to one atom of oxygen. A *water-right* is a legal claim to abstract or otherwise access water. You can float on water but not on a water-right. You can abrogate a water-right but not water. Chapter 9 is the only paper that addresses water-rights markets rather than water markets. It begins with pure theory on how urban actors' demand function for abstraction-rights intersects with farmers' supply function and thereby produces an equilibrium price and quantity traded. An account is given on how each function is determined. A number of real-life complications, such as part-sales and transaction costs, are examined and shown to be well-handled by neo-classical theory. But the empirical material suggests the modest relevance of the model to actual sales of water-rights where the absolutely predominant form of transaction is the bilateral deal. The conclusion is that, with respect to research method, fieldwork should be orientated to asymmetric power and information in the tradition of institutional economics.

1.10 FINAL REMARKS

I am convinced that, in human if not in financial terms, the most valuable areas of research in the 21st century will include humankind's understanding and management of water resources. Economists have a part to play here, alongside those who work in the fields of agriculture, development studies, engineering, environmental science, geography, hydrology and hydrogeology, law, planning, political science and sociology.

Of course, the greatest difficulty in getting to grips with water is that it slips through your fingers. Nevertheless, our work as researchers should be bold, honest, should respect the labour of disciplines other than our own, and should seek to capture the glittering diamond of method, theory, fieldwork and policy development. I believe that the century before us will witness water deficit on an unparalleled scale. I hope that this collection of papers helps colleagues in the global catchment to prepare for the challenges that will confront our species and the world-wide web of other life forms.

2

The regional water balance statement: a new tool for water resources planning

The king … proceeded to make use of the multitudes whom he had brought with him from the conquered countries … partly to dig the numerous canals with which the whole of Egypt is intersected … The king's object was to supply Nile water to the inhabitants of the towns situated in the mid-country, and not lying upon the river; for previously they had been obliged, after the subsidence of the floods, to drink a brackish water which they obtained from wells.

Herodotus c. 430 B.C.

2.1 INTRODUCTION

It is widely accepted that the effective planning of water resources becomes more urgent with each passing year. Some would argue that this need is driven by the increasing size and density of human populations at the catchment and urban scale. Others point out that exponential growth in economic output and consumption produces ever higher volumes of waste water (Lundqvist et al. 1985: 1). More recently, it is also asserted that global climate change will require every society to develop strategies capable of dealing with regional shifts in the mean and variance of hydrological variables such as precipitation.

In *Introduction to the Economics of Water Resources: An International Perspective* I have proposed that such planning should take place within a framework determined by the quest for a sustainable society, and that strategy for the water sector should be balanced by programmes on both the supply- and the demand-side (Merrett 1997: 187). Moreover, while our developing understanding of the hydrological cycle

by Stephen Merrett. Published by IWA Publishing, London, UK. ISBN: 1 84339 081 7.

provides the vital natural science input to strategy, it must be *the hydrosocial cycle* – illustrated in Figure 2.1 – which sources the language of supply and demand programmes. At this level, the key intellectual inputs come from politics, law, civil engineering, human geography, environmental studies and political economy.

The objective of this paper is ambitious. Derived from the hydrosocial cycle perspective on supply and demand, the paper seeks to develop a quantitative tool for water resource planning which within a decade could be used across the globe for strategy development in respect of abstracted fresh water for outstream uses. This tool is the regional water balance statement and its derivative, the change statement.

2.2 THE RULES OF THE GAME

The paradigm formulation of the regional water balance statement is presented in Table 2.1. Before looking at the individual entries in this template, some rules of the game are required. *First*, the statement relates to *any* defined geographic space; the term 'region' is used advisedly because of its inherent ambiguity. A region could be a continent, a country, a province, a catchment, an irrigation district, a city, a village, the site of a manufacturing firm such as a sugar mill, or, indeed, a water services company located in a defined area.

Secondly, the statement is always applicable to a defined time-period. For convenience of exposition it is assumed here that the time-period is the year 1999. Alternatively it could be for the 5 years from 1995 to 1999, for example, or for the month of August averaged over the 10-year period 1990–99. Regional water balance statements for past or present time are referred to as *baseline statements*; those for future time are *scenario statements*.

The statement has four columns: the first two are the categories of supply and the quantity supplied per unit period of time; the second two are the categories of use and the quantity used per unit period of time. The quantitative measure of flow will be chosen on pragmatic grounds; it is assumed in Table 2.1 that we refer to flows in megalitres per day, averaged over the year.

The boundary between supply flow and use flow should also be chosen on practical grounds – this paper is not intended to be a visionary text. Such a boundary might be the point where the user first possesses or has a right to the use of the water. Here, total use includes the leakage, evaporation and wastage which occurs on user properties, as well as beneficial use.

The calculation of flows in the regional water balance statement requires a procedure I shall call double-entry water accounting. The approach is taken from the double-entry book-keeping first developed in medieval Italy, and now universally used in modern financial accounting (Dyson 1994: 42–45). Within this routine, the company accountant has a large number of separate accounts such as a bank account, a creditors account, a debtors account, a cash account, a sales account and so on and so forth. Within the full set of a company's accounts, each transaction is recorded twice. In the account deemed to provide funds for the transaction, the transaction is recorded as a credit item. In the account deemed to receive the funds, the same transaction is recorded as a debit item. For example, the transaction of paying cash into the bank will be recorded as a credit in the cash account and as a debit in the bank account; the transaction of a cash receipt for the sale of goods is recorded as a debit

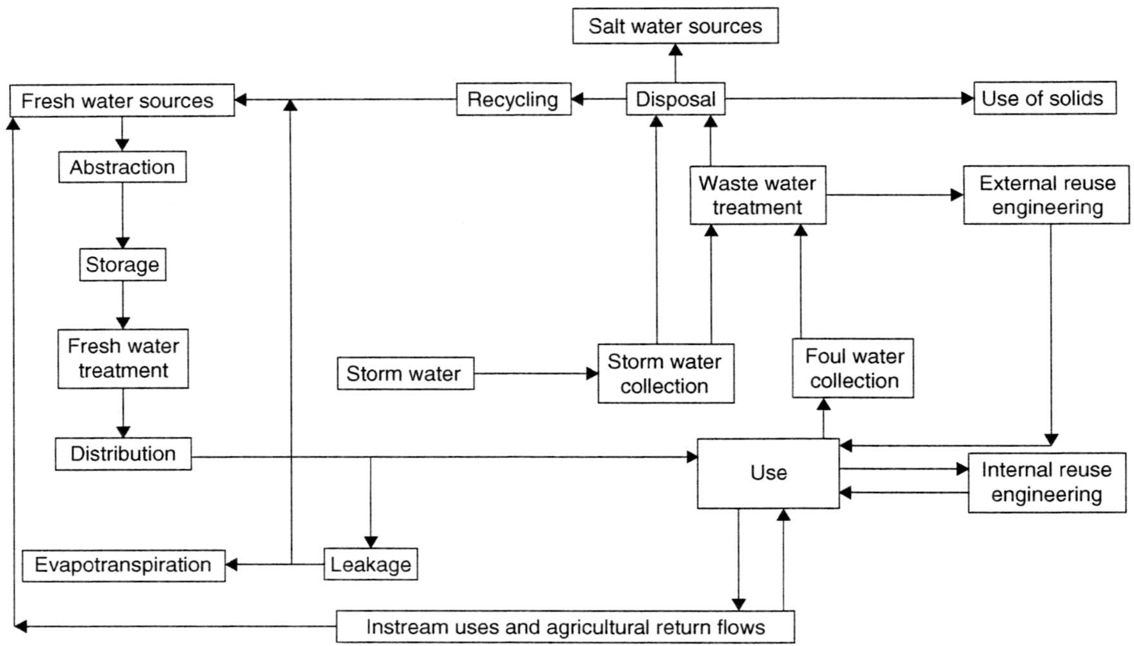

Figure 2.1 A simple model of the hydrosocial cycle.

Table 2.1 Regional water balance statement for the year 1999.

Categories of supply	Megalitres/day	Categories of use*	Megalitres/day
1. Rainwater collection	A	14. Households	S
2. Groundwater abstraction: first time through	B	15. Agriculture	T
3. Groundwater abstraction: recycled sources	C	16. Mining	U
4. Surface water abstraction: first time through	D	17. Manufacturing	V
5. Surface water abstraction: recycled sources	E	18. Public services	W
6. Desalination of salt or brackish waters	F	19. Commercial sectors	X
7. Import of water from another region	G	20. Instream applications	Y
8. Internal re-use of waste water	H	21. Other uses	Z
9. External re-use of waste water	J	**22. Total use**	$\boldsymbol{S + T \cdots + Y + Z}$
10. Less: supply leakage and evaporation	K		
11. Less: export of water to another region	L		
12. Fall or rise of volume of stored water	M		
13. Total net supply	$\boldsymbol{A + B + \cdots + J - K - L \pm M}$		

* Includes beneficial use, re-use volumes, and leakage, evaporation and wastage on user property.

item in the cash account and as a credit item in the sales account. In the process of 'balancing the books', because of this double-entry approach, the aggregate of all credit items in the separate accounts must equal the aggregate of all the debit items. Should this not happen, the accounts contain one or more errors of recording.

The application of financial accounting practice to the regional water balance statement must now be described. The statement is treated as a single ledger containing all the appropriate entries either as supply items on the left-hand side or as use items on the right-hand side. For any flow of water to be recorded in the statement, it must qualify as some hydrosocial category of input to the regional system. These input flows will be referred to as red molecule flows. In Table 2.1, flows 1–9 are red molecule flows. Such flows are parallel to the credit items described above and are entered on the left hand side of the water accounts ledger.

Once so entered, each molecule must be assigned to one of the hydrosocial output flows. These output flows are referred to as blue molecule flows. In Table 2.1, flows 10 and 11 and 14–21 are blue molecule flows. Such flows are parallel to the debit items described above and are entered either as negative values under the supply column or as positive values under the use column.

Using these two colours, red and blue, in writing and printing any specific statement has a heuristic value – water accounting throws up some puzzling questions. Because each red molecule from the supply side is re-entered as a blue molecule (negatively under supply or positively under use) total net supply is mathematically identical to total use. In the water accounts ledger, with comprehensive and accurate records, *the statement always balances*.

It goes almost without saying that it is only as a thought-experiment that we can identify every input molecule and trace it through to its metamorphosis as a blue molecule. One hundred per cent tracing is not the point here. The crux of the technique is that, in principle, every drop of water supplied in the region in 1999 can be allocated to either supply-side losses to the region or to regional use. A single set of molecules is being categorized in two different ways, on entrance to the set of accounts and on exit from it, and this is why the mathematical identity of total net supply and total use holds true.

2.3 THE SUPPLY CATEGORIES

Let us now consider in turn each of the categories of supply in Table 2.1. Category 1 is rainwater collection. This needs no gloss; it is found from the hamlets of the Caribbean island of Providencia to the site of the Millennium Dome in Greenwich. Rainwater collection always was and always will be with us.

Categories 2–5 refer to the abstraction of groundwater and surface water; each of these two sources is sub-divided into 'first time through' supply and 'recycled' supply. The use here of the term recycling requires clarification. The hydrosocial cycle of Figure 2.1 includes the collection of storm water and foul water and (after any treatment they receive) their routing back to rivers, lakes and aquifers. This and only this I call recycling. The term does not embrace disposal to the sea.

Where regional abstraction takes place downstream of recycling points, such abstraction flows can be decomposed into two parts: the abstraction of first-time water and the abstraction of recycled water. The quantitative ratio of these two flows can be taken to be equal to the ratio of fresh water to recycled water in the stream flow at the abstraction point. Table 2.1 incorporates the distinction by breaking down into two parts each of the red molecule abstraction flows. But note that the recycled flow *into* the fresh water source is neither a red nor a blue molecule flow. Recycled water is recorded in the statement only when it is abstracted.

In some regions, it may be right to ignore this break-down because it has no policy interest, no relevance to foreseeable infrastructural investment. But where sustainable water resource planning seeks to protect the hydrological cycle by the use of properly-treated recycled water, the distinction may be vital and measurement justified. Recycling adds to effective rainfall as a source of water for abstraction, as can be seen from the planning documentation of the Thames catchment (NRA 1994). The water flows of the Thames are among the most intensively used in the world.

Categories 6 and 7 add the two red molecule flows: desalination of salt or brackish water, and the import of water from another region. This is big-time supply-fix territory, this is the western USA, this is California. Also added, for the sake of symmetry in water transfers, is the blue molecule flow of water exported to another region, supply category 11.

In addition to recycling, Figure 2.1 also includes two further green loops: internal and external reuse. Internal reuse occurs when a household or a factory or any other organization reuses its own waste water. The water volume of internal reuse is set equal to each cubic metre of fresh water supplied to the user multiplied by the average number of times it is reused. External reuse occurs when the waste water of one organization or group of households is reused by a separate body, as in the reuse of treated waste water by agriculture. These loops are included in Table 2.1 as the red molecule flows of supply categories 8 and 9. *The same volumes are entered as blue molecule flows in the categories of use.* In this way, the fundamental mathematical identity is retained. Total net supply including reuse supply is equal to total use including its reuse flows.

My remarks above on the policy relevance of measuring recycled water apply also to reuse. Note that the distinction between recycled and reused water is that the former is water returned after its first use to river, lake and aquifer whilst the latter goes for reuse before disposal to fresh or salt water sinks.

Supply category 10 is (like 11) a blue molecule flow; these leakages and evaporation in the supply system between the points of abstraction and the supply/use boundary are deducted in the calculation of total net supply.

2.4 WATER STORAGE

Up to this point, the balancing of the water accounts ledger has derived from the notion that each and every molecule of water recorded as an input to the regional system in a given year, as part of a red molecule flow, is then recorded as an output from the system in that year as part of a blue molecule flow. The question then arises: does the existence of water storage infrastructures destroy the accounting balance? Our understanding of this issue will be strengthened if we imagine the region's storage capacity as composed of just three reservoirs, one black, one gold and one green.

The black reservoir is dedicated to the storage of water abstracted in 1999 that in the same year is distributed in its entirety to users or lost to leakage and evaporation.

Clearly, storage of these pass-through molecules in the black reservoir does not change the systemic balance for 1999.

The gold reservoir contains stored water abstracted during time-periods prior to 1999. These molecules are, so to speak, a gift from the past to the present. In 1999 some of the stored water is lost to supply leakage and net evaporation, some may be delivered as water exports to another region, and some is distributed to users. The fall in the quantity of water stored in the gold reservoir during 1999 is expressed in megalitres per day and is deemed to be a red molecule flow. Once again, the identity of total net supply and total use is maintained.

The green reservoir is dedicated to the receipt of water abstracted during 1999, which will be stored for distribution from 2000 onwards, a gift from the present to the future. The increase in the volume stored in the green reservoir during 1999 is expressed in megalitres per day and is deemed to be a blue molecule flow. This flow precisely matches the abstraction flow pumped to the reservoir and, for the third time, the mathematical identity holds.

In practice, of course, each reservoir in a real regional system combines the functions of all three reservoirs described above. What we observe is only the net outcome of the component processes, that is, either no change in 1999 in the volume of stored water, or a fall or a rise. Thus, with respect to the value of M in Table 2.1, no change in the total volume of stored water gives a value of zero; a fall is expressed at its daily rate and is recorded as a red molecule flow; and a rise in storage is expressed at its daily rate and is recorded as a blue molecule, negatively-valued flow. The analysis here of reservoir storage applies with equal force to aquifer storage and recovery.

2.5 THE USE CATEGORIES

The use categories 14–21 of Table 2.1 can be swiftly dealt with. The categorization is designed to group together similar types of use so that use forecasting is facilitated. It has already been noted that the term use includes beneficial use, reuse, and the leakage, evaporation and wastage occurring on user properties. In planning for any real region categories 14–21 would be redesigned in the manner most effective for strategy development in that area.

Only category 20 requires further comment. It is already stated above that the regional water balance statement is a tool for strategy development in respect of abstracted fresh water for outstream uses. The statement does not, in general, address instream uses of water for nature conservation, leisure pursuits, fishing and navigation. These uses are not of a kind that makes double-entry water accounting relevant. However, instream flows are sometimes supplemented by abstraction. For example, in England and Wales British Waterways pumps groundwater to maintain the necessary volumetric flows in its canals without which navigation and boating is hindered. Similarly, the Royal Society for the Protection of Birds in the UK may use water service company water supplies for application to the wetlands it manages in order to maintain their environmental quality at times when rainfall and river flows are deficient. Where red molecule flows are allocated in this way, these applications must be recorded in category 20 in order to retain the necessary equivalence of total net supply and total use.

2.6 THE CHANGE STATEMENT

Regional water balance statements, as already indicated above, may be baseline statements for the past or present or they may be scenario statements for the future. When the baseline and the scenario statement have the same structure, we can subtract the entries of the former from the entries of the latter to give a regional water balance *change* statement. This result is illustrated in Table 2.2, where lower-case letters are used to indicate we are dealing with *differences* in values. Ten rules for the change statement are worth setting out; they are all derived from a single arithmetic rule of subtraction of baseline (positive, zero and negative) values from scenario (positive, zero and negative) values.

- Where the value of a cell is unchanged between the baseline and the scenario statements, the change value is zero.
- In the case of each of the red molecule flows from rainwater collection to external reuse, if the value in the scenario year exceeds that of the baseline year, the change value is positive.
- In the case of each of the red molecule flows from rainwater collection to external reuse, if the value in the scenario year falls short of that of the baseline year, the change value is negative.
- In the case of each of the blue molecule flows from household use to other uses, if the value in the scenario year exceeds that of the baseline year, the change value is positive.
- In the case of each of the blue molecule flows from household use to other uses, if the value in the scenario year falls short of that of the baseline year, the change value is negative.
- In the case of each of the supply-side blue molecule flows (supply leakage/evaporation and export of water to another region), if the *absolute* value in the scenario year exceeds that of the baseline year, the change value is negative. For example, if we have a shift from a base year value of 7 Ml/d exported water to a scenario year value of 11 Ml/d, the value of the letter *l* is −4 Ml/d.
- In the case of each of the supply-side blue molecule flows (supply leakage/evaporation and export of water to another region), if the *absolute* value in the scenario year falls short of that of the baseline year, the change value is positive.
- In the case of stored water, the value of m is positive when a fall in the scenario year exceeds a fall in the baseline year or when a rise in the baseline year is succeeded by a fall in the scenario year.
- In the case of stored water, the value of *m* is negative when a fall in the scenario year falls short of a fall in the baseline year or when a fall in the baseline year is succeeded by a rise in the scenario year.

From these rules we can see that any single lower-case value may be positive, zero or negative. In the change statement the total change in net supply is equal to the sum of entries *a* to *m*. Similarly, the total change in use is equal to the sum of entries *s* to *z*.

A final rule of great importance can be established. Since total net supply and total use in the baseline year are identically equal, and since total net supply and total use in the scenario year are identically equal, it follows like the night the day that the change between the 2 years are identically equal. So, in the regional water balance change statement, Merrett's law states:

- Total change in net supply is mathematically identical with total change in use.

Table 2.2 A regional water balance change statement for the 10-year period between 1999 and 2009.

Categories of supply	Changes in megalitres/day	Categories of use*	Changes in megalitres/day
1. Rainwater collection	a	14. Households	s
2. Groundwater abstraction: first time through	b	15. Agriculture	t
3. Groundwater abstraction: recycled sources	c	16. Mining	u
4. Surface water abstraction: first time through	d	17. Manufacturing	v
5. Surface water abstraction: recycled sources	e	18. Public services	w
6. Desalination of salt or brackish waters	f	19. Commercial sectors	x
7. Import of water from another region	g	20. Instream applications	y
8. Internal re-use of waste water	h	21. Other uses	z
9. External re-use of waste water	j	**22. Total Change in Use**	$\boldsymbol{s + t + \cdots + y + z}$
10. Less: supply leakage and evaporation	k		
11. Less: export of water to another region	l		
12. Fall or rise of volume of stored water	m		
13. Total Changes in Net Supply	$\boldsymbol{a + b + \ldots + l + m}$		

* Includes beneficial use, re-use volumes, and leakage, evaporation and wastage on user property.

I have suggested that the change statement is derived as the difference between the baseline and the scenario statements. In planning practice, it may be common for a change statement to be added to a baseline statement to produce a scenario statement.

2.7 THE USES OF REGIONAL STATEMENTS

The time has arrived to outline the potential value of these statements and to propose why they may become a routine tool of water resource planning in a wide range of countries over the next decade.

The baseline statement can be prepared for any geographic area and provides a comprehensive, synoptic account of the scale and composition of the supply sources of water and its uses in that region. The approach is pragmatic: one can foresee an evolution in statements for a specific region from the simplest set of water accounts to more complex ones, as policy interests drive the expansion of the categories deployed and the accuracy of their measurement. The work process to make the baseline calculations will stimulate new directions for research and will deepen planners' understanding of the region's hydrosocial cycle within its hydrological context. The four-column approach gives equal weight both to supply-side and demand-side information and describes them in the language of a unified conceptual framework. The recycling and reuse entries will strengthen our understanding of the cyclical components of water resource planning, make them more visible. The fundamental identity of total net supply and total use provides a consistency check on individual calculations, akin to the trial balance procedure of company accountants (Merrett 1997: 115–24).

The change statement, the planning tool *par excellence*, is applicable to short-, medium- and long-term strategy, to comparison of typical years past and future, and it clarifies the quantitative choices that a regional government or catchment agency will be forced to envisage in a future year of severe drought, or as a result of climatic change in hydrological variables. The change statements can provide comprehensive, synoptic, transparent alternative scenarios within a region for infrastructural investment and for shifts in the scale and pattern of uses. It is in this way that it handles the uncertainty that, as Peter Rogers writes, is a major issue in water resources planning (1991: 1–8), as well as side-stepping the inadequacy of project-by-project planning (Rogers 1992: 1). The change statement does not evaluate a scenario, this is the work of environmental scientists, economists and others, but it does impose a consistency check on each case, because of the fundamental identity of total change in net supply with total change in use.

The approach to a regional water balance statement presented here can be contrasted with current, orthodox methods. For example, Philip Anderson (1997) adopts the conventional approach in recent work on Scotland. In his study the demand side is based on social processes while the resource side is derived from natural processes in the form of hydrological yields, and where the resource and demand totals do not, in fact, balance. As a result of this yields approach, his overview of the Scottish situation tabulates only one of the 12 supply-side variables listed in Table 2.1, that is, leakage.

A second example of the current orthodoxy is the work of Miloradov & Marjanovic (1998): *Guidelines for conducting water resources assessment*. It is worth noting that this publication is one of UNESCO's studies and reports in hydrology and was written as a contribution to the International Hydrological Programme. Not surprisingly, the authors' approach to flows, stocks and balances is essentially hydrological, not hydrosocial. Their Chapter 4 is entitled *Basic theoretical principles for processing a water resources balance*. It suggests that the core component of any water resources master plan is the water resources management balance (WRMB). However, in a graphic illustrating the water resources balance approach, the supply-side variables are predominantly hydrological and the uses of water are not represented at all. Moreover the terms of the general WRMB equation for a given area in a given time-period embraces a mixture of hydrological variables (such as rainfall and aquifer flows) and engineering variables (such as artificial water conduit flows and discharges by water users). The general equation and the specific equations derived from it simply do not seek to report the balance between how human society in a given area gains access to its water and how it uses it. So the WRMB, however necessary, is strikingly different from the regional water balance statement presented in this paper.

My third and last example of the received wisdom is the *Assessment of water resources and water availability in the world* authored by I.A. Shiklomanov and published by the World Meteorological Organization (WMO) in 1997. Chapters 1–4 are entirely hydrological in their orientation to resource availability, with a special stress on river runoff. Chapter 5 (*Water resources use*) and Chapter 6 (*Water availability and water resources deficit in the world*) contain valuable synoptic material on patterns of water consumption but the supply-side concept never moves beyond either the generic term 'water withdrawal' or the two-fold breakdown of supply into surface water and groundwater sources. Once again the approach is unaware of the richness of the concept of supply when it is located within the hydrosocial rather than the hydrological paradigm, with the huge relevance this has to catchment planning; and at no point does the WMO text set up a regional water balance statement conceived as a mathematical identity.

2.8 CONCLUSIONS

I have argued that a regional water balance statement of the type presented in Table 2.1 and the change statement of Table 2.2 make a radical break with existing approaches, in fact complement them, and that the new tool has a strong relevance to water resource planning practice. But it is a mistake to gild the lily. The limitations of the statement must also be recognized: it is a quantitative technique that embodies no hydrological or meteorological variables; for all practical purposes it does not address instream uses; and it makes no reference to water quality. Thus the regional water balance statement should be seen only as one of the many, *interdependent* approaches and techniques necessary for effective water resource management.

The material in this chapter originally appeared in: The regional water balance statement: a new tool for water resources planning. *Water International*, 24(3): 268–274, 1999.

3

Integrated water resources management and the hydrosocial balance

To see a world in a grain of sand...

William Blake (1803)

3.1 INTRODUCTION

Integrated water resources management (IWRM) seeks to integrate in many different ways: the social sciences with the natural sciences, planning with implementation, and groundwater with surface water, to mention just three such ways. The first objective of this paper is briefly to introduce a planning method, the hydrosocial balance, and to show how it integrates: (i) outstream water quantities supplied and used, and (ii) the present with the future. The paper then shows how the hydrosocial balance can be developed to integrate water qualities with water quantities. The paper's third and final objective is to apply this management tool to a case-study from the island of Jersey in the English Channel.

3.2 A NEW IMPLEMENT FOR THE IWRM TOOLBOX

This section recapitulates the main features of a new water resources management tool, the hydrosocial balance, developed over several years. It is a tool that until now has been limited to the analysis only of water quantities (Merrett 1997: 15–22; 1999: 268–74; 2002: 148–53).

The hydrosocial balance is a quantitative water resources planning method applicable in principle to any space with a defined boundary. This might be a house on the coast of Andalucía, Wembley football stadium, the island of Gorgona, the catchment of the Yangtse, the State of California, or the continent of Africa. Such spaces are referred to here generically as 'regions' and are not required to be related to traditional 'catchment' areas. The hydrosocial balance's principal distinctiveness from the 'water balance' familiar to hydrologists is that the former incorporates (almost without exception) only outstream, hydrosocial flows, and never the hydrological flows of precipitation, groundwater recharge, run-off and rivers. The water flows that it places at the centre of analysis and measurement are those directly created by human society; the idea of a hydrosocial balance is derived from the concept of the hydrosocial cycle (Merrett 1997: 6–7). A hydrosocial flow represents a human activity. So the hydrosocial balance, composed as it is of many hydrosocial flows, is understood primarily through the social sciences. In contrast, the hydrological balance represents natural flows and is understood primarily through the natural sciences.

To summarize, hydrological flows are of a type that exists in a state of nature, prior to the recent appearance of *Homo sapiens*. Hydrosocial flows, in contrast, are specific to human society. In recent centuries these two types of flow have become ever more powerfully interdependent with the growth of world population and its economic activities of production and consumption. The water balance of an area and that area's hydrosocial balance should be estimated separately prior to considering their quantitative and qualitative interdependence.

The generic form of the hydrosocial balance for a specified region is given in Table 3.1. A baseline balance is for a past time-period, such as the year 2001. A scenario balance is for a future time-period such as the year 2007. The shift in the quantity in millions of cubic metres (Mcm) of any one category of supply or use between the baseline year and the scenario year can be represented both as an absolute change, vide column 4 of Table 3.1, as well as an annual rate of growth or decline, *vide* column 5.

The baseline balance provides a comprehensive, synoptic account both of the scale and composition of the supply sources of water as well as their use in the region it covers. Where measurement is accurate and comprehensive, the total net supply is always equal to total use. Scenario balances provide options for the future, based on the forecast need for outstream water in different uses and the possible allocation conflicts that may be foreseen. Once again, total net supply must be planned to equal total use. The absolute difference of supply, and of use, between the base year and any specific scenario year, together with the associated annual rate of change, provide the basic input to the planning of infrastructural investment, capital financing and demand management.

3.3 THE BRIDGE BETWEEN QUANTITY AND QUALITY

The previous section focused entirely on quantity. In this section a bridge is built that links quantity to quality. Table 3.1, for a given base year or scenario year, has more than a dozen supply-side and demand-side flows. All these flows, plus the post-use flows of waste water (including irrigation drainage), can be reclassified into the five groups set out in Table 3.2. Note that in a region where there is no treatment of the water supply, or of waste water, one has only groups one, three and four: the supply flow, the use flow and the waste water flow.

Table 3.1 The hydrosocial balance for a specified region in a base year and a scenario year (in millions of cubic metres, mcm).

	Base year	Scenario year	Scenario year minus base year (+ or −)	Annual compound rate of growth from the base year to the scenario year(%)(+ or −)
Categories of Supply				
Rainwater collection	A1	A2	A2 − A1	Ga
Groundwater abstraction	B1	B2	B2 − B1	Gb
Surface water abstraction	C1	C2	C2 − C1	Gc
Desalination	D1	D2	D2 − D1	Gd
Import of water from other regions	E1	E2	E2 − E1	Ge
Internal reuse of wastewater	F1	F2	F2 − F1	Gf
External reuse of wastewater	G1	G2	G2 − G1	Gg
Total Gross Supply	H1	H2	H2 − H1	Gh
Supply leakage and evaporation	−J1	−J2	(−J2) − (−J1)	Gj
Export of water to other regions	−K1	−K2	(−K2) − (−K1)	Gk
Fall (+) or rise (−) in volume of water abstracted and stored	±L1	±L2	(±L2) − (±L1)	–
Total Net Supply	M1	M2	M2 − M1	Gm
Categories of Use				
Households	S1	S2	S2 − S1	Gs
Agriculture	T1	T2	T2 − T1	Gt
Mining	U1	U2	U2 − U1	Gu
Manufacturing	V1	V2	V2 − V1	Gv
Public services	W1	W2	W2 − W1	Gw
Private services	X1	X2	X2 − X1	Gx
Other uses	Y1	Y2	Y2 − Y1	Gy
Total Use	Z1	Z2	Z2 − Z1	Gz

Note: Gj and Gk are calculated using absolute values of leakage and exports. Gl is not calculated because of the possible change of sign.
Source: Adapted from Merrett (2002a), Tables 7.1 and 7.2.

Table 3.2 Flow types in the hydrosocial cycle.

Supply-side flows		Use-flows	Waste water flows	
1	2	3	4	5
Prior to treatment	After treatment	At the point of use	Prior to treatment	After treatment

Note. The supply-side flows include supply leakage. The waste water flows include irrigation drainage. Different levels of treatment produce different quality products at different costs with different implications for users and the environment.

For any of the flows of Table 3.2, its complex quality can be assessed provided that a water institution has the skilled professionals and the laboratories to carry out the necessary analysis. At the most general level, a flow's quality can be measured by applying four criteria to samples taken from the flow:

1. The individual inorganics present in the samples (such as arsenic, lead and zinc).
2. The individual organics present (such as atrazine, malathion and 2,4-D).
3. The microbiological content of the samples (in terms such as faecal coliforms, pathogenic staphylococci and salmonella).
4. Other indicator measures (such as biochemical oxygen demand, total suspended solids and pH).

However, if one considers the hundreds of individual characteristics that can be generated by these four criteria, it would require a prodigious hydrochemical infrastructure to process comprehensively even a single sample from a single flow. So measurement must always be targeted, principally by considering the 'fit for purpose' needs of the analysis. That is to say, one reviews what the water flow under assessment is to be used for, or to which location it is to be discharged. So, at the most elementary level, if the water that is sampled has to meet drinking water standards, as it will be pumped to domestic premises, then the water quality assessment is quite different from that for water one plans to discharge to coastal waters. Moreover, the two flows will be governed by different legislation and standards.

In summary, the bridge linking the quantities of the hydrosocial balance to their qualities is built in the following manner. One recognizes first that the quantities of the hydrosocial balance fall into the flow types of Table 3.2; second that each flow's quality can be assessed in terms of the four criteria listed above; and third that the specification of the assessment should be based on the resources available to carry it out, the legislative requirements to meet prescribed standards, and the fit-for-purpose requirements of the hydrosocial balance flows themselves.

This suggests a new term is required. When one has a cross-tabulation for a specific hydrosocial flow with: (i) rows that refer to that flow's qualitative characteristics, and (ii) columns setting out the number of samples made and the measured concentration per litre or measured value of each characteristic, this will be referred to as a *quality matrix*.

Up to this point the text is at a high order of generality. Sections 3.4–3.7 record an attempt to apply this analytic framework, or meta-theory, to the island of Jersey. Beginning with the geographical and hydrological background, the paper moves on to cover the supply-side of the hydrosocial balance, then the use of water in Jersey, and finally the complex issues of water quality. The case-study benefits from recent hydrogeological publications of great quality, as well as a series of interviews with key personnel.

3.4 THE PHYSICAL GEOGRAPHY AND HYDROLOGY OF JERSEY

The island of Jersey is located in the English Channel, some 60 km north of the French coastal town of St Malo. It has an area of 117 km^2 and in 2001 had a resident population of about 85,000 persons, giving a density of 726 persons/km^2. As Robins (2000: 2–3) points out, the island comprises a plateau lying at an elevation of 60–120 m divided by a series of valleys running from north to south. The bedrock beneath Jersey mostly consists of ancient metamorphosed shales of Precambrian age intruded by younger igneous rocks. Unconsolidated deposits of Quaternary age partially cover the bedrock, whilst post-glacial peat, alluvium and sand are also present in valleys and low-lying coastal areas.

Westerly and south-westerly winds bring moisture from the Atlantic. Long-term average rainfall for the island is 877 mm, annually varying from 600 to 1100 mm in the past decade. Mean annual potential evapotranspiration ranges from 648 to 784 mm. The 136-year monthly average rainfall shows that the bulk of precipitation occurs in the 7 months September–March (Jersey New Waterworks Company 2002: 4).

Flow along the valley streams is north to south and, in the absence of a dry winter, takes place throughout the year. Groundwater discharges both to these streams and at the coast. During prolonged dry weather, saline intrusion may occur locally. Robins writes (2000: 8): 'There is a notable variation in run-off and infiltration from year to year. Poor winter rains had a marked effect on values for the notorious dry years 1975/76, and the periods 1989 to 1992 and 1995/96. These dry years are significant, as it is these years of water stress that the Island needs to be able to cope with in terms of surface water storage capacity and conjunctive abstraction of groundwater.'

Robins also shows that the island's base flow index has an estimated value of 0.58, indicating that 58% of streamflow is derived from the groundwater baseflow (Robins 2000: Table 1; Blackie et al. 1996). This quantity will be of importance in the discussion of water quality in Section 3.7.

3.5 WATER SUPPLY IN JERSEY

At this point one can begin to review the principal actors in Jersey that supply water and that use it. As this is done, the jigsaw of Jersey's hydrosocial balance is assembled in Table 3.3.

There can be no question that the principal supply-side provider is the Jersey New Waterworks Company (JNWWC). This is the island's only water utility. However, waste water collection, treatment and discharge are the responsibility of the States of Jersey's Public Services Department (PSD).

The JNWWC is a water undertaking incorporated in 1882 and its principal activities are the supply and treatment of water for domestic use and other purposes on the island. The States of Jersey, the island's government, holds 100% of the issued 'A'

Table 3.3 The hydrosocial balance of the states of Jersey in 2001.

Row	Categories of supply	Mcm
3	Surface water and groundwater abstraction by the JNWWC	6.2
4	Desalination by the JNWWC	1.1
5	Groundwater abstraction by farmers, households, etc.	3.6
6	Surface water abstraction by farmers	not known
7	Rainwater collection net of evaporation	<0.1
8	Internal reuse	<0.1
9	External reuse	<0.1
10	Imports from other regions	0.0
11	**Total Gross Supply**	**10.9**
12	Less: exports to other regions	0.0
13	Less: supply leakage and evaporation	−0.7
14	Change in volume of water abstracted and stored	<0.1
15	**Total Net Supply**	**10.2**
16	**Categories of Use**	Mcm
17	Sourced by the JNWWC	
18	Households	4.5
19	Agriculture	<0.1
20	Manufacturing	<0.1
21	Public services	0.6
22	Private services	1.3
23	Other uses	<0.1
24	**Sub-Total**	**6.5**
25	Sourced by groundwater pumped by farmers, households, etc.	
26	Households	0.9
27	Agriculture	1.4
28	Industry	0.1
29	Hotels and hospitals	0.1
30	Leisure	0.8
31	Other	0.3
32	**Sub-Total**	**3.6**
33	Irrigation water use sourced by farmers' surface water abstraction	not known
34	Irrigation water use sourced by rainwater collection	<0.1
35	Error term	+0.1
36	**Total Use**	**10.2**

ordinary shares. However, the utility acts like a private company. It has a desalination plant, six reservoirs, an extensive network of raw water mains, two treatment plants and a distribution infrastructure to the island's users. The total length of the trunk mains is 70 km and of the service mains 415 km. In 2001 the company had a turnover of £11.0 million, it employed 112 persons and the balance sheet shows that its fixed assets, excluding landholdings, were valued at £47.7 million (JNWWC 2002).

Of the annual volume of water abstracted by the JNWWC, the overwhelming share is from its surface water reservoirs. Total abstraction in 2001 was 6.2 Mcm to which one should add 1.1 Mcm output from the desalination plant at La Rosière. This is shown in Table 3.3, rows 3 and 4. Note that, for management purposes, there is a good deal of switching of surface water between reservoirs, but from the perspective of the island as a whole, this can be ignored in calculating the abstraction total. The low reliance on groundwater as a primary source of supply, in contrast to surface water, may be because the prime costs of bulk surface water abstraction are lower than for groundwater, given that the main aquifers are of only poor to moderate transmissivity. Another explanation is that the company believes groundwater quality to be inferior to that of surface water in Jersey (see Section 3.7 below).

If the JNWWC is the principal supply-side provider, the second most significant supply source is a group made up of individual farmers and households. Agriculture's needs for water are met largely from supply points on the farms themselves. These are either boreholes or ponds fed by small streams. The financial costs of pumping from these boreholes and ponds are borne by the farmer. At the present time, boreholes are neither registered nor licensed nor metered, nor is any abstraction charge levied by the States. The same holds true for abstraction from farm ponds; in this case farmers usually have mobile diesel engines to pump the water from pond to field or to farm buildings (Vint, pers. comm.).

About 15% of the island's resident households are unconnected to the JNWWC network and about 17% have no link to the PSD's sewerage. There is a large overlap between these two groups of the more isolated households. Families off the water supply network use boreholes and wells for their domestic supplies. Once again these individual supply points are unregistered, unlicensed, unmetered and there is no charge for their abstraction. One can say that access to the aquifer by these individual farmers and households is outside the sphere of the States' management of the island's water resources.

This unmeasured and unmanaged characteristic of surface water and groundwater abstraction by most farmers and a minority of households raises obvious difficulties in estimating Jersey's hydrosocial balance. However, estimates of groundwater abstraction and springflow interception were made for the dry years 1989–91, when borehole meters were installed following the State of Emergency powers granted during the 1989 drought (Robins 2000: 15). The number of supply sources at that time were: 500 in agriculture, 4000 in the domestic sector, 50 in the leisure sector and about 80 others. Robins suggests that the average use of groundwater remained generally stable during the 1990s and is likely to be little changed 'to this day'. This gives an annual rate of ground water abstraction by private actors of approximately 3.6 Mcm in Table 3.3, row 5.

Rainwater collection is now considered. It is known that some householders not connected to the mains supply collect water on their roofs and route it to a cistern. The scale of this is believed to be small. More important is the greenhouse sector. This includes polytunnels and glasshouses. Some farmers have very large glasshouses for the commercial production of flowers, vegetables and soft fruit. They capture run-off from the roofs by channelling it along the roof-gutters and through downpipes to water tanks. The Jersey Department of Agriculture and Fisheries (DAF) suggest that the total area of glasshouse roofs is some 51,000 m^2. Net of evaporative loss, rainfall on these roofs was 537 mm in 2001, giving rainwater collection in this case of about 27,000 m^3; this also is recorded in Table 3.3, row 7. The evaporative loss is low because rainfall is channelled quickly and directly to the collection tanks.

Internal and external reuse in Table 3.3 now needs to be considered. Reuse refers to water resource flows that, after abstraction, are first used in the domestic, industrial or irrigation sectors and then are used for a second time or, indeed, time and time again prior to recycling back to lake, river and aquifer (Merrett 2002a: 8). Internal reuse occurs when the repeated use is by the same institution where the water was first used. This is familiar in the industrial sector, and increasingly in the domestic sector where grey water is reused to irrigate gardens and to flush toilets. External reuse takes place when the repeated use is by a different institution, such as the use of waste water from the cities for the irrigation of farmland; the cases of Ghana and Israel can be cited here. The importance of internal and external reuse to the hydro-economist is that it multiplies the productivity of a given volume of water appropriated as a base flow. The water quality of the effluent reused and the degree of treatment (if any) that it undergoes are critical issues. In response to questions on this subject to a number of water resource professionals in Jersey there was a common answer: cases of reuse can be found but the aggregate scale on the island is negligibly small. The most common case is one of internal reuse: households that use their kitchen water to irrigate their gardens.

The last entry under total gross supply in Table 3.3 is the bulk import of water to Jersey. This simply does not take place.

One can now estimate the total gross supply; the figure is 10.9 Mcm for the year 2001. However, this total excludes the volume of one non-negligible entry for which no estimate exists: surface water abstraction by farmers from the ponds and pools on their land that are fed by small streams. In future the Water Resources (Jersey) Law that has recently been drafted will address this.

To move from total gross supply to total net supply, three adjustments have to be made: for exports, losses and storage change. Exports of bulk water by Jersey to other regions, as with imports, are zero. However, supply-side leakage and evaporation of abstracted water between the point of abstraction up to the boundary of users' properties is of real importance. The JNWWC has a number of staff who work to keep losses down to an economic level. The company estimates that in 2001 their losses were 6% of gross supply, an extraordinarily low figure by British and world standards. In the complete absence of other data, the same percentage is also used here for groundwater abstraction by farmers, households and others. This gives an entry in Table 3.3, row 13 of −0.7 Mcm.

Lastly, row 14 in Table 3.3 refers to changes in the volume of water abstracted and stored. The logical place of this category in the hydrosocial balance is described at length elsewhere (Merrett 2002a: 150–1). It is sufficient to say here that the entry refers only to changes in the volume of stored water that has already been captured, appropriated or abstracted in some way. It does not refer to hydrological changes in the volume of water in a lake or behind a barrage resulting from variation in a catchment's river flows. In the case of Jersey, the volume of abstracted water in storage in this sense is quite small at any point in time, so that changes in its volume over the course of a year are negligibly small and can be ignored. The end result is that, for the time being, the best estimate of total net supply in Jersey for the year 2001 is 10.2 million cubic metres.

3.6 WATER USE IN JERSEY

This section addresses water use on Jersey in the baseline year of 2001. The use data in Table 3.3 can be seen to fall into two principal blocks: uses sourced by the JNWWC and those sourced by private abstraction of groundwater. Examination of the table's categories of use and the volumes known to be used lead to a clear conclusion: 88% of the known total is made up by just three groups of users. In descending order these are: households, private services and agriculture.

Use by households is the largest category by far, totalling 5.4 Mcm. In the year 2001, the resident population of Jersey was about 85,000. There are individual houses and villages all over the island but the largest population concentration is on the south coast, particularly in Jersey's capital, St Helier. Dividing use by resident population gives a figure of 174 litres per head per day (lhd) for domestic use. This is rather high compared with the UK average of 147 lhd, which in any case is (mistakenly) inclusive of supply-side losses (Environment Agency 2002). Of the 5.4 million total, 4.1 Mcm is used by JNWWC customers who are not metered. Their domestic water supply charges are based on the capital value of their property. So there exists no price signal to encourage demand management in these cases. This may explain, partially at least, the high domestic use average.

Private services rank second among the users of water in Table 3.3. The JNWWC's water demand staff provide a useful breakdown of this category. It includes, in descending order of volume: hotels and guest houses; offices, banks and shops; public houses; restaurants; sports facilities; stores, garages and car parks; clubs, parish halls and day centres; laundries; and hair and beauty salons. The two dominant economic contributors to Jersey's gross domestic product are the financial services sector and tourism; this explains the salience of private services as a user of water. At the height of the summer there are around 35,000 tourists on the island; this generates a strong summer surge in demand. All commercial properties are metered. However, there is no metering of tourists' and visitors' individual water use within their hotels and lodging houses.

Even excluding the irrigation water pumped from on-farm ponds, agriculture is the third largest user of water. Uses include net irrigation requirements, livestock watering, cleansing of farm buildings and machinery, and product washing. Traditionally the Jersey economy has been regarded as quintessentially agricultural (Frigot 2001). Even in 2001, farmland constituted 49.6% of the island's area, with 352 farmholdings. The principal outputs from the sector are 'Jersey Royal' potatoes, tomatoes, and the dairy products of the island's 4550 Jersey cows and heifers (DAF 2002). Jersey soils are inherently low in organic matter and in most cases multi-cropping requires the addition of fertilizer.

Returning to Table 3.3, one sees sub-totals of 5.4 Mcm for household use (rows 18 and 26), 2.1 Mcm for private services (22 and 30) and 1.4 Mcm for agriculture (19 and 27). Adding in the remaining uses specified, as well as an error term of +0.1 Mcm to offset the difference between estimated total net supply and total recorded use, gives total use of 10.2 Mcm.

3.7 WATER QUALITY IN JERSEY

The water quality of Jersey is a large subject. Here the discussion is limited to five hydrosocial flows selected because of their particular interest for water resources management. These five flows are the irrigation returns of farm use, the abstraction of groundwater by households unconnected to the JNWWC's water supply, the domestic waste water from households unconnected to the PSD's sewerage, the

JNWWC's distribution to households, private and public services, etc. and the PSD's discharges to coastal waters. Quality matrices will be deployed below for the last two of these five flows, to illustrate the link of quality to quantity.

3.7.1 Irrigation returns

The water requirements of Jersey's crops are met both by rainfall and by irrigation, as is the case with the water needs of its parks and golf courses. Part of this joint flow is consumed as evapotranspiration. The remainder flows by gravity as field drainage, either as run-off to local streams or by percolating down to the aquifer. At present, no water quality analysis is directly made of agricultural drainage in the island on a regular basis. However, a number of studies exist that are relevant to linkage between farmland drainage and groundwater quality. Robins & Smedley (1998) have presented the evidence in great detail. Jersey's farmers do not stint in their use of fertilizers, particularly for early potatoes. These include the traditional 20–10–10 NPK product as well as livestock slurry, the manure of grazing cattle and, now on a much-reduced scale in comparison with the past, *vraic* – the local term for seaweed collected from the beaches. These inputs contain *inter alia* nitrates, phosphorus and potassium. Robins (2000: 26), reporting on work by Lott et al. (1999), states: '... the continued zealous use of nitrate fertilizer and application of farm slurry remains largely above recommended levels for the UK ...'. Chilton & Bird (1994) did carry out research specifically into the pore water below farmers' fields. They noted correlations with fertilizer use not only of nitrates but also of the other common fertilizer ions of ammonium, chloride and sulphate. Since Lott's report, the Department of Agriculture and Fisheries, in a personal communication, writes that there is increasing evidence that a majority of the farming industry has now adopted changed practices.

Farmers also use herbicides and fungicides, and their residues and metabolites exist in farmland drainage. Until it was banned in Jersey in 1998, the pesticide chlorthal dimethyl was an ingredient in pesticides applied to brassicas and soft fruits. A principal degradation product, the metabolite chlorthal, is environmentally persistent. Matthews & Carter (1999) modelled its movements and concluded that by 2002 farmland drainage would in all cases have concentrations less than the critical level of 0.1 micrograms per litre. However, by October 2002 it was still well above the critical level in surface water and groundwater (G. Jackson, pers. comm.).

Certainly there exists a Water Pollution Law (PSD 2000) to safeguard surface water and groundwater from contamination. This is targeted principally at point sources but Articles 12–16 can tackle diffuse pollution issues. The DAF also has an agriculture–environment scheme that aims at the better management of slurry, reduced fertilizer use and increased care in the use of biocides. The Water Pollution Law will also be used to introduce water management areas and these areas may themselves stimulate action to reduce the contamination that originates from farmland drainage. Such drainage is said to be the origin of some 75% of the nitrate contamination of the island's groundwater (JNWWC 2002: 5).

In the case of this first flow, farmland drainage sourced by irrigation returns and rainfall, there is a strong case for wishing to examine its current quality matrix, but no such matrix appears to exist.

3.7.2 Groundwater abstracted by households for their domestic needs

We have already seen that about 15% of the island's households have no water supply connection to the JNWWC. They rely on their own wells and boreholes and are thereby exposed to the raw state of the aquifer and thereby to the diffuse contamination of groundwater by farming practice. There are believed to be approximately 5200 domestic premises dependent on groundwater sources. Robins and Smedley write: 'Under the UK Environment Agency's classification, all aquifers in Jersey would be classified as "highly vulnerable" because of the importance of fracturing (by-pass flow), lack of impermeable cover and the shallow depth to the water table' (Robins & Smedley 1998: 17).

To reduce their exposure to groundwater contaminants some households have invested in private treatment systems. The public health aspects of domestic borehole supply lies with the Department of Environmental Health. A sample of 52 boreholes and wells across the island are monitored in May and November of each year for inorganics, 16 of which are further analysed for pesticides and microbiological contamination.

3.7.3 Discharges by households of their own waste water

This is the second case of households unconnected to the island's dominant infrastructures. It is said that about 17% of households have no access to the PSD's sewerage. The special position of this group lies in the quality of its domestic waste water discharges. In fact, the PSD reports (2000: 1) 'Sewage pollution of water from malfunctioning private drainage systems is common...'. Robins & Smedley (1998: 13) suggest that the average daily discharge per household is about 600 litres. An analysis of the Val de la Mare catchment's nitrogen export coefficients by Lott et al. (1999) showed that 10% of the nitrogen was sourced by domestic water from septic tanks and soakaways, which was as much as the contamination by livestock.

The Water Pollution (Jersey) Law 2000 requires that owners of septic tanks, tight tanks, soakaways or private sewage treatment plants have a discharge permit only where such discharges result in or are likely to result in the contamination of controlled waters. Gass et al. (1996: 45) recommend that all soakaways and septic tanks should be taken out of use and sewerage extended to the whole island population. However, they do not say why effective private treatment by households should be eliminated, nor do they estimate the economic costs of extending the PSD's sewerage network.

3.7.4 The JNWWC's supply to its customers

In earlier sections of this paper it has been shown that the JNWWC supplies two-thirds of Jersey's total gross supply, excluding surface abstraction by farmers. The qualitative assessment of its treated water supply is detailed and appears in Table 3.4. The quality matrices published by the JNWWC (2002: 28–30) cover altogether 33 characteristics of the treated water supply, two of which are for microbiological standards. In the case of the other three quality categories – individual inorganics, individual organics and other measures – four or five examples from each are selected, including all those cases where one or more samples exceed the maximum admissible concentration (MAC).

Table 3.4 is a baseline statement of quantities in respect of 16 separate qualitative characteristics of a hydrosocial flow. Just as the hydrosocial balance can be assigned a scenario balance, so too can a quality matrix be assigned a scenario matrix. The parallel between the hydrosocial balance and the quality matrix is strong here and

Table 3.4 A quality matrix for the JNWWC'S treated water supply in Jersey in 2001.

	Maximum admissible concentration or value (MAC)[2]	Concentration			Number of samples taken	% of samples > MAC
		Minimum	Mean	Maximum		
Individual Inorganics						
Nitrate	50 milligrams NO_3/litre	28.5	46.7	69.0	157	31
Nitrite	0.1 milligrams NO_2/litre	0.001	0.037	0.219	155	12
Chloride	400 milligrams Cl/litre	54	73	90	155	0
Manganese	50 micrograms Mn/litre	<20	<20	65	157	<1
Lead	50 micrograms Pb/litre	<1	5	53	74	1
Individual Organics						
Atrazine	0.1 micrograms/litre	<0.01	<0.01	0.013	10	0
Simazine	0.1 micrograms/litre	<0.01	<0.01	0.012	10	0
Cyanazine	0.1 micrograms/litre	<0.01	<0.02	0.12	52	1
Mecoprop	0.1 micrograms/litre	<0.01	<0.01	0.02	50	0
Dalapon	0.1 micrograms/litre	<0.01	<0.01	0.02	9	0
Microbiological Standards						
Total coliforms[1]					523	<1
Faecal coliforms[1]					523	0
Other Measures						
pH	6.5–9.5	7.2	7.4	8.3	229	0
Turbidity (suspended solids)	4 N.T.U.	0.08	0.27	1.5	154	0
Colour	20 Hazen units	<0.5	4.3	5.0	155	0
Dissolved solids	1500 milligrams/litre	230	389	485	153	0

Notes: [1] Zone 1 – East. Random consumer taps and fixed points. [2] These all appear to be EU values but the source does not make this clear.
Source: JNWWC (2002): 28–30.

shows that, in an IWRM context, both the balance and the matrix can be treated in parallel ways, and can have a baseline table calculated and a scenario table set as a target for water resource managers. The principal difference between balance and matrix is that a region has only a single baseline hydrosocial balance but it has a baseline quality matrix for every hydrosocial flow that the water management authority wishes to measure and to plan for.

In this paper only one of the rows of Table 3.4 is examined, that for nitrate. Nitrate's principal point of entry into Jersey's hydrological and hydrosocial flows has already been discussed above in Section 3.7.1. The downstream outcome is that in 2001 some 31% of the treated water supply to the island's population exceeded the European Union's MAC of 50 milligrams of NO_3 per litre (mg/l). During the first 9 months of 2002, there was no case of the MAC being exceeded.

Here it should be pointed out that the measure of nitrogen present in water because of nitrate contamination may take two forms:

1. The measured mass of the complete nitrate ion in the water, referred to as milligrams of NO_3/litre.
2. The mass of nitrogen in the water that is locked into the measured mass of the nitrate ion, referred to as milligrams of NO_3-N/litre.

For a given sample, the ratio of (1) to (2) is 4.42 to 1. Therefore, it is absolutely essential that any document or discussion of nitrate contamination makes clear which of these two alternative measures is being used. Here the data and the European Union (EU) MAC are expressed in terms of (1) above.

Why in the past did the government of Jersey permit nitrate contamination of the island's drinking water supplies? Politically, it was because government was not willing to introduce measures that would be contrary to the farming interest. Farmers, in fact, made up more than their proportionate share of Jersey's political class. This is now changing fast. Some of the measures introduced or about to be introduced, which promise an amelioration of the current contamination of groundwater, have already been referred to above.

With specific reference to samples in excess of the nitrate MAC (see above), the formal defence is made on the basis of an argument that cannot lightly be dismissed. In the past, World Health Organization standards for drinking water were to 'recommend' less than 50 mg/l and to regard as 'acceptable' 50–100 mg/l. But in 1996 the WHO fell into line with the EU's criterion, first introduced in 1980, of 50 mg/l. The States of Jersey are not a member of the EU and are under no legal or constitutional obligation to comply with its water quality directives. Moreover, to do so in respect of 6.2 million cubic metres of water distributed per year (Table 3.3) would be costly. The JNWWC already desalinates water at some expense in order to blend it with surface water abstracted so that the mean concentration of NO_3 drops below 50 mg/l (see Table 3.4). The health argument advanced by the States is based on the views of medical experts in the UK. The Joint Committee on the Medical Aspects of Water Quality appointed by the Department of the Environment and the Department of Health and Social Services stated that 'There is no compelling evidence to suggest that significant risks to health are encountered when water containing between 50 and 100 mg/l nitrate is supplied to the public' (Gass et al. 1996: 35–36; JNWWC 2002: 29; Robins 2000: 15–17).

In Table 3.4 the MAC is also exceeded with respect to nitrite and, by a small margin, with manganese, lead, cyanazine, and total coliforms in the east of the island. It is noteworthy that the EU, on health evidence, plans to reduce the MAC for lead from

50 down to 25 micrograms per litre in December 2003, and further down to 10 micrograms by 2013 (N. Walton, pers. comm.). Particularly in the case of nitrate, nitrite, lead and the metabolite chlorthal, it seems clear that Jersey's water resource managers need to agree on a scenario quality matrix for the treated water supply. A recent consulting contract let by the States of Jersey under the title Water Quality Objectives may lead to just such a matrix.

3.7.5 The PSD's discharges to coastal waters

As has already been mentioned, waste water collection, treatment and discharge are the responsibility of the States of Jersey's Public Services Department. The volume of such discharges (including stormwater) in 2001 was some 12 Mcm. There are two treatment plants, one at Bellozanne in the south of the island and a second, small satellite plant recently commissioned on the north coast at Bonne Nuit Bay. Bellozanne consists of two inlet 3 mm step screens, fat/grit separation at the inlet works, four primary clarifiers, an activated sludge treatment plant, 12 final clarifiers and a tertiary ultraviolet disinfection system. Discharges from the two plants are regulated within the terms and conditions of a discharge certificate that requires regular sampling to monitor compliance levels for the effluent leaving the works.

The conditions in the discharge certificate for the Bellozanne sewage treatment works (STW) are based on the requirements of the EU's Urban Waste Water Treatment Directive (European Commission 1991) but prior to 2003 were less stringent than those set out in that Directive. The STW at the end of 2002 was undergoing conversion to a 'Pegazur' system in order to reduce nitrogen inputs into the nearby St. Aubin's Bay. In addition, the ultraviolet-applied dose will be raised from a minimum of 14 mW s/cm^2 in 2002 up to 24 mW s/cm^2 from July 2003. The current Jersey maximum admissible concentrations of waste water contaminants are given in Table 3.5 as

Table 3.5 A quality matrix for the PSD's waste water discharges to Jersey's coastal waters in 2002.

	Concentration						
	Measure	MAC in			Actual 2002		
		2002	2003	2004	Minimum	Mean	Maximum
Biochemical oxygen demand	milligrams per litre	50	25	25	6.3	8.2	13.4
Chemical oxygen demand	milligrams per litre	250	125	125	21.0	46.0	64.0
Suspended solids	milligrams per litre	150	45	35	11.5	18.5	29.8
Total nitrogen	milligrams per litre	none	20	10 or 15	not measured	not measured	not measured

Note: Actual 2002 data are based on monthly averages for January–October.
Source: PSD (pers. comm.).

well as the minimum, mean and maximum actual values achieved for the period January–October 2002. From 2003, the PSD will add total nitrogen to biochemical oxygen demand, chemical oxygen demand and suspended solids as the criteria of the quality matrix for waste water and stormwater discharges.

3.8 CONCLUSIONS

With respect to the outstream flows of any defined region, whether or not it is a catchment, this paper has a central thesis: integrated water resources management is the planned transformation over time in the flow quantities of the region's hydrosocial balance alongside the purposive change of the quality matrix of each policy-priority flow. It is hoped that this paper indicates the usefulness of this approach to the management of outstream flows. What remains to be done is to draw some lessons from the Jersey case-study that will benefit planners who in the future follow this methodological path.

3.8.1 The hydrosocial balance

Jersey is a 'region' in the language of this paper, i.e. a space with a defined boundary. The fact that it is composed of many small water catchments creates no problem. The year selected as the baseline is the most recent for which data are available. The scenario year does not appear in this case-study because, at present, the island's water resource planners do not use this method and so there is no scenario year. As a general rule, the scenario year should not be so close to the baseline year that it gives too little time for plan implementation nor so distant that the projection becomes an implausibility in the misty distance. A 5–7-year plan period is about right. The hydrological assumptions for the scenario year should reflect the region's long-term average. Since these assumptions are likely to be different from the actual conditions in the baseline year, the changes between say 2001 and 2007 (column 4 of Table 3.1) will, in part, reflect the difference.

The row structure of the hydrosocial balance is always the same: the components of supply making up total gross supply, the three entries for losses and storage change, and the categories of use. The column structure of base year, scenario year, absolute difference and rate of change are also a standard feature.

With respect to supply, the Jersey data of Table 3.3 are structured principally by the actors responsible. This has advantages in policy formulation. The main difficulty is the unavailability of 2001 data on groundwater abstraction by private persons and surface water abstraction by farmers. This leads to a policy recommendation: that the island's authorities should begin to estimate these data on a regular basis, either by sample metering or by an interview-based survey. Constructing the hydrosocial balance for a region always leads to proposals for additional data collection. The same argument holds true for rainwater harvesting and for reuse, sources often neglected that may deserve policy initiatives. If these categories are not entered into the hydrosocial balance because the values are not known, they are invisible to the planner.

Turning to exports, losses and storage change, the Jersey story appears simple. There are no exports. Storage change is negligible (see Section 3.5). Loss estimation and reduction are a priority for the main supplier, the JNWWC, but no loss estimate exists for private abstraction; the water utility's 6% datum is used in this case.

With respect to use, the data are available separately for the two dominant sources of supply, i.e. the JNWWC and private actors. From both an analytic and a policy point of view this has its advantages. Unfortunately, the classification of types of use differs between the sources, creating an adding-up problem. The researcher's choice of supply and use categories should always reflect the particularities of the region in which the work takes place as well as its planning focus. It was surprising to discover that 'private services' has the second largest recorded use after 'households'. This is explained by the vital role of tourism and financial services in the island as well as by the fact that (outside the polytunnel and glasshouse sector) irrigation is supplementary to the island's (877 mm average) rainfall. The main weakness of the use data is that for private actors it is an estimate based on a sample last made in 1989–91. The lack of registration, licensing and metering of private groundwater abstraction in Jersey deserves to be reviewed by the States' government. In fact the Water Resources (Jersey) Law to control abstraction and impoundment will be presented to the States in 2003.

The final comment on Table 3.3 is that an error term of 0.1 Mcm was added to total use to give equality with the total net supply figure of 10.2 Mcm. Properly defined and accurately and comprehensively measured, total net supply is mathematically identical to total use. In practice, any real-life calculation will always contain a disparity as water measurements are never exact. Unfortunately, with respect to Table 3.3, one cannot conclude that the error is only of the order of 1% (0.1/10.2). This is because the table is sure to contain mutually-cancelling errors.

3.8.2 Quality matrices

The principal methodological objective of this paper is to integrate quality with quantity, that is, to bring together quality matrices with the hydrosocial balance. In deploying matrices in the field the biggest challenge is to specify which of the potentially innumerable flows should be measured, given the high overhead and prime costs of quality assessment. Key guidelines to such a choice are:

- The importance of any specific flow's quality in understanding how other downstream flows are polluted.
- The impact of any specific flow's quality on the environmental health of the region's population.
- The significance of a specific flow's quality on the natural environment.
- The requirements of the governing legislation, standards and guidelines for that flow.

Using Tables 3.2 and 3.3, just five flows were selected for discussion in the Jersey case-study. The first is irrigation returns with its recognized wide externalities. Here it is immediately clear that the hydrosocial flow has to be combined with farmland drainage sourced by rainfall; the two are not separable. Much is available on the sources of contamination of farm drainage and the relative importance of such contamination island-wide. However, there appears to be no up-to-date quality matrix for this specific flow. Jersey's planners may wish to take action here, particularly in the light of the EU's nitrate-sensitive areas designations.

The second selected flow is groundwater abstraction by households for their domestic needs, because they are unconnected to the JNWWC's water supply network. These families are exposed to the aquifer in its polluted condition. This flow is monitored twice per year for its quality.

The third flow is households' discharge of their own waste water via septic tanks and soakaways and concerns households unconnected to the PSD's sewerage network.

They may thereby pollute groundwater, for example with ammonia, detergent residues and faecal coli.

The fourth and fifth flows are the JNWWC's water supply and the PSD's discharge of waste water and stormwater to the sea. These are the dominant hydrosocial flows of the island and are discussed at greater length in this paper. In both cases the quality matrices are detailed and informative. Currently, the main policy issues are the nitrate content of the water utility's output and the move towards more stringent maximum admissible concentrations for the sewage treatment works at Bellozanne.

With these comments on the practical construction of the hydrosocial balance and the quality matrices of some of its flows, this paper is now complete. It is hoped that the research may be of practical use to the people and institutions of Jersey as well as suggesting, to professional colleagues, new ways of integrating water resources management in respect of the relationship between the quantity and quality of outstream, hydrosocial flows.

The material in this chapter originally appeared in: Integrated water resources management and the hydrosocial balance. *Water International*, 29(2): 148–157, 2004.

4

Farm-level drought management: an irrigation case-study from the UK

There is special providence in the fall of a sparrow. If it be now, 'tis not to come; if it be not to come, it will be now; if it be not now, yet it will come. The readiness is all.

William Shakespeare, *Hamlet*

4.1 INTRODUCTION

The farmer's world is encompassed by risk and no risk is more devastating than a prolonged drought. Yet, just as risk can be managed, drought can be planned for. The purpose of this paper is to use a case-study from the Anglian Region in England to show the preparations being taken on a specific farm for the region's next severe drought, and how risk analysis permitted an evaluation of two alternative, farm-level drought management strategies.

The broader context of the case-study is the introduction of a Water Bill into the UK Parliament in February 2003. The Bill, once enacted, promises a more effective water resources planning system and specifically creates an obligation on water companies to develop drought plans for submission to the Environment Agency (House of Lords 2003 clauses 60–2). This Agency is the statutory body in England and Wales for strategic water resources management. A key pressure to which it seeks to respond is climate change. As van Hofwegen and Svendsen (2000: 9–10) write:

> The gradual warming of the earth, 1.0°C in the past 50 years, is leading to glacial recession, declining snow cover, and rising sea levels. Precipitation patterns

are likely to alter, reducing water availability in some regions and increasing it in others. Increased variability in precipitation patterns will accompany this shift with a huge impact on both irrigated and non-irrigated agriculture. Precipitation patterns will include a greater proportion of extreme events, leading to higher and more frequent flooding and lower dry season flows in rivers.

Within this 'broader context' referred to above, it is important in the UK to understand how farmers themselves will manage drought risk in a water company's drought plan area.

The structure of the paper is straightforward. It begins with a brief description of the Anglian Region itself, followed by a general overview of the farm's activities. Thereafter an 'infrastructural strategy' for drought years is compared with an 'informational strategy' based on long-term rainfall data; the alternatives are evaluated in terms of relative cost. The paper ends with a brief summary and conclusions.

4.2 THE ANGLIAN REGION

The Environment Agency is divided into eight regions, one of which is Anglian. This region stretches from the Humber Estuary in the north to the River Thames in the south, from the Norfolk coast in the east to Northampton in the west. It covers 27,000 km^2 and five million people live in the area. The region has extensive, sparsely populated rural areas, particularly in the north and east, a long coastline and interspersed urban centres around which industry has developed. As communication links continue to improve (especially with London) it is likely to be one of the highest population growth areas in the country.

The Anglian Region has 58% of the most productive agricultural land in England and Wales and agriculture has considerable influence on the rural economy and communities. The sustainable use of water is crucial given the competing demands for it, particularly because this region is the driest in the UK. Low flows in late summer and increased demand especially for agriculture and garden use are likely, say the Environment Agency, to cause stress (www.environment-agency.gov.uk/) In some summers, irrigation can make up 50% of total use.

The second principal actor in water resources management in the region is Anglian Water Services Limited (AWS), which serves the needs of over five million industrial, commercial and domestic consumers (www.anglianwater.co.uk/). It is this company that will have to submit its drought plan to the Environment Agency once the Water Bill is enacted by the UK Parliament. In fact already in the spring of 2003 the company had submitted an *Anglian Water Services Drought Plan March 2003* to the Environment Agency.

AWS is the regional utility and its principal activities are the provision of drinking water and the treatment of waste water. The company employs 3700 full-time equivalent persons and in 2002 had a turnover of £724 million (AWS 2002). Its principal infrastructural assets are mains and sewers, impounding and pumped raw water storage reservoirs, dams, sludge pipelines and sea outfalls. The company's capital investment plan in the 5 years through 2004/5 is equal to £1.5 billion and covers further improvements to drinking water, bathing and river water quality, as well as network improvements to maintain serviceability and meet new demands.

4.3 SILVER BIRCHES plc

In this section are reviewed the activities of the agribusiness at the centre of this case-study, an enterprise referred to here as 'Silver Birches plc'. The farm is located in the Ely area of the Anglian Region on soils typical of the district around Cambridge (Hodge & Seale 1966).

Silver Birches plc is 57 hectares in size. The product is container trees and the business is the largest such nursery in Europe. Maximum capacity in terms of the stock at any point in time is 130,000 trees ranging from 8–10 cm girth in 45 litre pots through to 25–30 cm girth in 300 litre pots. The company sells trees of about 330 different varieties in its unique, easy-to-handle, white containers. The amenity market is its customer and this includes universities, hospitals, retail, residential and office developments, town centres, golf clubs and football grounds.

In order to understand the farm's access to and use of irrigation water, the planning concept of the hydrosocial balance is set out in Table 4.1. This requires considerable adaptation for the specific case of Silver Birches plc. Beginning with the positive cate-

Table 4.1 The hydrosocial balance for a specified area in a base year.

	Base year volumes (mcm)
Categories of Supply (Positive)	
Rainwater collection	a
Groundwater abstraction	b
Surface water abstraction	c
Desalination	d
Imports of water from other areas	e
Internal reuse of wastewater	f
External reuse of wastewater	g
Net fall in water abstracted and stored	h
Total Gross Supply	**j**
Categories of Supply (Negative)	
Supply leakage and evaporation	k
Exports of water to other regions	l
Net rise in water abstracted and stored	m
Total Negative Values	**n**
Total Net Supply	**j − n**
Categories of Use	
Households	s
Agriculture	t
Mining	u
Manufacturing	v
Public services	w
Private services	x
Other uses	y
Total Use	**z**

Note. A net fall (rise) in the storage of abstracted water has a positive (negative) value.
Source: Adapted from Merrett (2002a) Tables 7.1 and 7.2.

Table 4.2 The hydrosocial balance for Silver Birches plc in 2002.

Row		Quantity (cubic metres)
5	**Categories of Supply (Positive)**	
6	Rainwater collection	0
7	Groundwater abstraction	0
8	Surface water abstraction	0
9	Desalination	0
10	Mains water	39,315
11	Internal reuse of wastewater	0
12	External reuse of wastewater	0
13	Net fall in water abstracted and stored	Negligible
14	**Total Gross Supply**	**39,315**
15		
16	**Categories of Supply (Negative)**	
17	Supply leakage and evaporation	400
18	Exports of water to other regions	0
19	Net rise in water abstracted and stored	Negligible
20	**Total Negative Values**	**400**
21		
22	**Total Net Supply**	**38,915**
23		
24	**Categories of Use**	
25	Buildings	350
26	Irrigation	38,565
27	**Total Use**	**38,915**

gories of supply, Table 4.2 shows that in the year 2002 there was no rainwater collection, no groundwater or surface water abstraction, no desalination and no reuse of waste water. Moreover, with respect to Table 4.1's categories of net fall or rise in storage, at Silver Birches plc on-farm storage of purchased mains water is sufficient only for 48 hours use. This is necessary in case of a brief failure in the mains supply, but any change in storage over the course of a whole year is negligibly small. Two days storage is only 215 m^3. The absolutely dominant supply source is mains water purchases from Anglian Water Services. AWS has a 12-inch main running close to the farm and off this a link is made to a 3-inch and a 6-inch pipe; the latter runs like a spinal cord down the centre of the farm.

With respect to the negative categories of supply, on-farm losses of mains water are low: 1% is the current estimate and this is entered in row 17 of Table 4.2. Losses are monitored using the 3-inch meter at the connection to the utility's main, as well as the meters and pressure gauges inside each of the farm's four pump-houses. In most cases on-farm supply pipes are electrowelded and Silver Birches plc has two skilled industrial plumbers on its staff who are on-call on a 24/7 basis. There are no exports of water. As a result, total net supply was about 39,000 m^3.

In principle, total net supply is mathematically identical to total use (Merrett 2002: 150). As Table 4.2 shows, use is simply divided into buildings and irrigation. Farm management has not developed a split of use into different species of tree or different locations of tree or different container sizes. Such a division of use is not (at present) considered to have any value in the management process. Ninety-eight per cent of containers at the farm are divided roughly equally between the 45- and 85-litre sizes. The enterprise deploys a Netafim pressure compensating non-leakage (CNL)

Table 4.3 Silver Birches plc: monthly Anglian Water Services supply 2002 (cubic metres).

Row	Month billed	Volume supplied	Number of metered days	Volume per metered day	Volume for adjusted month
	Irrigation				
5	January	22	35	1	19
6	February	35	29	1	37
7	March	248	34	7	222
8	April	3426	28	122	3720
9	May	4328	28	155	4699
10	June	8449	35	241	7339
11	July	6127	28	219	6652
12	August	5379	28	192	5840
13	September	7771	36	216	6562
14	October	2975	27	110	3350
15	November	132	28	5	143
16	December (estimate)	77	30	3	78
17	Total	38,969	366		
18	Average			106	3247
19					
20	**Central Services**				
21	January–December	346			
22					
23	**Grand Total**	**39,315**			

Note: The number of metered days in column 3 is the number of days that the end-of-month invoice covered. Volume for the adjusted month applies the volume per day of column 4 to a standard month of 30.4 days.

drip system to each separate tree where each container has one manifold, one CNL dripper, microtubes and four arrow-stakes. Averaged over the whole stock, each tree received 1.12 litres per day in the farming year September 2001 to August 2002, when the mid-year stock equalled 94,000 trees, equal to about 2 years' sales. The variation in the rate of irrigation over the calendar year 2002 is given in Table 4.3. The volume of water delivered to the trees in the 5 months May–September is 80% of the total. An interesting statistic can be derived from these data, once we realize that the average surface area of the two dominant container-types is 0.159 m^2. At 1.12 litres/day, the average container consumes 2600 mm/year of irrigation water.

4.4 DROUGHT MANAGEMENT: THE INFRASTRUCTURAL STRATEGY

Silver Birches plc has grown rapidly in the past 14 years to attain its European pre-eminence as a supplier of container trees. But by the turn of the millennium the managing director (MD) had become increasingly concerned about the impact a severe and

Table 4.4 Supply planning at Silver Birches plc with an 18,000 m^3 safeguard against drought.

Year	Stock and flow (m^3)	September	October	November	December	January	February	March	April	May	June	July	August
Year 1 2003–4	Abstraction to reservoir by end of month	0	0	0	0	12,000	12,000	12,000	0	0	0	0	0
	Irrigation supply from non-drought storage	0	0	0	0	0	40	230	3720	4700	7340	1970	0
	Irrigation supply from Anglian Water	6570	3350	150	80	20	0	0	0	0	0	4690	5840
	Total irrigation supply	6570	3350	150	80	20	40	230	3720	4700	7340	6660	5840
	Drought storage by end of month	0	0	0	0	12,000	18,000	18,000	18,000	18,000	18,000	18,000	18,000
	Non-drought storage by end of month	0	0	0	0	0	5960	17,730	14,010	9310	1970	0	0
	Total storage	0	0	0	0	12,000	23,960	35,730	32,010	27,310	19,970	18,000	18,000

Year 2 2004–5	Abstraction to reservoir by end of month	0	0	7200	7200	3850	40	230	0	0	0	0	0
	Irrigation supply from non-drought storage	0	0	150	80	20	40	230	3720	4700	7340	2240	0
	Irrigation supply from Anglian Water	6570	3350	0	0	0	0	0	0	0	0	4420	5840
	Total irrigation supply	6570	3350	150	80	20	40	230	3720	4700	7340	6660	5840
	Drought storage by end of month	18,000	18,000	18,000	18,000	18,000	18,000	18,000	18,000	18,000	18,000	18,000	18,000
	Non-drought storage by end of month	0	0	7050	14,170	18,000	18,000	18,000	14,280	9580	2240	0	0
	Total storage	18,000	18,000	25,050	32,170	36,000	36,000	36,000	32,280	27,580	20,240	18,000	18,000
Year 3 2005–6 to 2022–23	As for 2004–5												

extended drought might have on the business. A young tree in a 45- or 85-litre container rapidly becomes stressed without irrigation, so much so that in the absence of water for 15 days it is irreparably damaged.

MD's concern is not *directly* linked to a regional drought because, as we have seen, at Silver Birches plc it is mains water that sources the soil moisture required by the growing trees. However, the risk exists that a severe drought would lead AWS to cut its supplies to the farm. In a conflict over access to scarce water, the order of sectors in the UK in terms of relative political strength is probably households, public services, industry, private services and, finally, the agricultural sector. MD's judgement is that the National Farmers Union in the Anglian Region does not have the muscle adequately to defend the agricultural sector from mains supply reductions.

In these circumstances MD realized that conjunctive supply could play a role in farm-level drought management. The Anglian Region is famously flat and land drainage is a matter of great importance (Merrett 2002a: 105–12, Talbot & Whiteman 1996). As a result the landscape is criss-crossed with innumerable unlined, surface drains. One such drain makes up the farm's northern boundary. MD secured an abstraction licence for this drain from the Environment Agency. It gives Silver Birches plc the right to abstract during the five winter months from 1 November to 31 March. The duration of the right is from May 2002 through March 2027. The licensed means of abstraction is a centrifugal pump of a maximum output not exceeding 29 litres/second. The maximum quantity permitted to be abstracted is 100 m^3/hour, 1000 m^3/day and 36,000 m^3 during the five winter months. Low flow provisions exist to protect the environment of the downstream South Level system. These are that the abstraction right is suspended when the combined flow of the Ely Ouse river and the cut-off channel at the Denver Sluices does not exceed 318,226 m^3/day (1 November–28 February) or 113,652 m^3/day (1 March–31 March).

The abstraction flow will go straight into a winter storage reservoir on the farm. The contract for this has now been signed and the reservoir should be ready to receive the off-take water from January 2004. It will have a storage capacity of 36,000 m^3. The new infrastructure will combine the reservoir, pump station and sand filtration.

Table 4.4 models the relative role of the winter storage supply and the mains supply in meeting the business's crop water requirements over the 20 farming years September 2003 through August 2023. This is here called the infrastructural strategy. Key assumptions in developing the model are:

- Monthly crop water requirements are equal to those of the adjusted monthly volume from Table 4.3.
- Abstraction during the winter months is spread equally over the months when it is permitted (in 2003–4 it begins only in January when the reservoir has been completed).
- Abstraction to the reservoir does not take place when the reservoir is full.
- Half the reservoir's capacity (18,000 m^3) is held back as an irrigation source should there be a cut-off in the mains supply. This is 'drought storage'.
- Irrigation supply from the reservoir begins only after the reservoir has reached its drought storage level.
- Irrigation supply from AWS is used only when there is no non-drought storage in the reservoir.
- The final assumption is that, in fact, there turns out to be no drought so severe over the twenty years that the 'drought storage' is required.

4.5 DROUGHT MANAGEMENT: THE INFORMATIONAL STRATEGY

There are two problems with the infrastructural strategy. First, it permanently dedicates half the capacity of the winter storage reservoir to drought storage when the severe and prolonged drought event that it protects against is extremely rare. Second, it reduces the irrigation supply that can be sourced from winter storage and therefore increases the irrigation supply sourced from AWS. The additional prime cost (marginal cost) of the AWS supply is £0.59/m^3 whereas from winter storage it is only £0.15/m^3. In the UK farmers pay only £0.15 per litre of diesel fuel.

Table 4.5 models the informational strategy. Its key assumptions are:

- Monthly crop water requirements are equal to those of the adjusted monthly volume from Table 4.3.
- Abstraction during the winter months is spread equally over the months when it is permitted (in 2003–4 it begins only in January when the reservoir has been completed).
- Abstraction to the reservoir does not take place when the reservoir is full.
- There is no provision in this model for drought storage.
- Irrigation supply from AWS is used only when there is no storage in the reservoir.
- The final assumption is that, in fact, there turns out to be no drought so severe over the 20 years that the 'drought storage' is required.

No delivery contracts or other legal mechanisms invalidate the "key assumptions" referred to above. The central difference between Tables 4.4 and 4.5 is of course that in Table 4.5 the whole of winter storage is applied for irrigation, thereby reducing purchases from AWS. Table 4.6 shows these volumetric differences and their cost implications.

The total excess of water purchases from AWS with the infrastructural strategy in comparison with the informational strategy is 347,000 m^3 over the 20 years 2003/4–2022/23. In terms of expenditure this makes the infrastructural strategy more expensive by a sum of £153,000. Discounting the stream of differential expenditure at the interest rate charged on Silver Birches plc's overdraft – base rate plus 2% i.e. 5.75% – we have a *discounted* sum of £88,000. In financial terms there is no doubt of the relative attraction of the informational strategy.

However, the reader will surely have noted a drawback of the informational strategy: thus far it has no provision for drought planning! In the remainder of this section it is shown how such planning *is* possible and why this leads me to call the second approach the 'informational strategy'.

We have already seen that low rainfall currently provides no direct threat to Silver Birches plc's farming practice. Virtually all of its crop water requirements are met from the mains supply. The drought risk takes the form of a possible reduction in the AWS supply. The firm's MD believes that this would most likely be during a dry summer that succeeded a dry winter, each of great severity.

So it made sense to review rainfall records for the area, kindly provided by the Environment Agency for the Isleham Pumping Station. The calendar year data show that the driest year since 1963 was in 1996 with total rainfall equal to 391 mm compared to the 1963–2002 average of 548 mm. *Yet in 1996 AWS imposed no restrictions of any kind on the mains supply to its customers.*

Table 4.5 Supply planning at Silver Birches plc with no fixed safeguard against drought.

Year	Stock and flow (m^3)	September	October	November	December	January	February	March	April	May	June	July	August
Year 1 2003–4	Abstraction to reservoir by end of month	0	0	0	0	12,000	12,000	12,000	0	0	0	0	0
	Irrigation supply from non-drought storage	0	0	0	0	20	40	230	3720	4700	7340	6660	5840
	Irrigation supply from Anglian Water	6570	3350	150	80	0	0	0	0	0	0	0	0
	Total irrigation supply	6570	3350	150	80	20	40	230	3720	4700	7340	6660	5840
	Drought storage by end of month	0	0	0	0	0	0	0	0	0	0	0	0
	Non-drought storage by end of month	0	0	0	0	11,980	23,940	35,710	31,990	27,290	19,950	13,290	7450
	Total storage	0	0	0	0	11,980	23,940	35,710	31,990	27,290	19,950	13,290	7450
Year 2 2004–5	Abstraction to reservoir by end of month	0	0	7200	7200	7200	7200	7200	0	0	0	0	0
	Irrigation supply from non-drought storage	6570	880	150	80	20	40	230	3720	4700	7340	6660	5840
	Irrigation supply from Anglian Water	0	2470	0	0	0	0	0	0	0	0	0	0

	Total irrigation supply	6570	3350	150	80	20	40	230	3720	4700	7340	6660	5840
	Drought storage by end of month	0	0	0	0	0	0	0	0	0	0	0	0
	Non-drought storage by end of month	880	0	7050	14,170	21,350	28,510	35,480	31,760	27,060	19,720	13,060	7220
	Total storage	880	0	7050	14,170	21,350	28,510	35,480	31,760	27,060	19,720	13,060	7220
Year 3 2005–6	Abstraction to reservoir by end of month	0	0	7200	7200	7200	7200	7200	0	0	0	0	0
	Irrigation supply from non-drought storage	6570	650	150	80	20	40	230	3720	4700	7340	6660	5840
	Irrigation supply from Anglian Water	0	2700	0	0	0	0	0	0	0	0	0	0
	Total irrigation supply	6570	3350	150	80	20	40	230	3720	4700	7340	6660	5840
	Drought storage by end of month	0	0	0	0	0	0	0	0	0	0	0	0
	Non-drought storage by end of month	650	0	7050	14,170	21,350	28,510	35,480	31,760	27,060	19,720	13,060	7220
	Total storage	650	0	7050	14,170	21,350	28,510	35,480	31,760	27,060	19,720	13,060	7220
Year 4 2006–7 to 2022–23	As for 2005–6												

Table 4.6 Purchases of irrigation water from Anglian Water Services under two alternative strategies.

Year	Irrigation water purchases	September	October	November	December	January	February	March	April	May	June	July	August
Year 1 2003–4	Infrastructure strategy (m^3)	6570	3350	150	80	20	0	0	0	0	0	4690	5840
	Information strategy (m^3)	6570	3350	150	80	0	0	0	0	0	0	0	0
	Marginal cost excess of AWS supply over winter storage supply (£/m^3)	0.44	0.44	0.44	0.44	0.44	0.44	0.44	0.44	0.44	0.44	0.44	0.44
	Cost savings of information strategy (£)	0	0	0	0	9	0	0	0	0	0	2064	2570
Year 2 2004–5	Infrastructure strategy (m^3)	6570	3350	0	0	0	0	0	0	0	0	4420	5840
	Information strategy (m^3)	0	2470	0	0	0	0	0	0	0	0	0	0
	Marginal cost excess of AWS supply over winter storage supply (£/m^3)	0.44	0.44	0.44	0.44	0.44	0.44	0.44	0.44	0.44	0.44	0.44	0.44
	Cost savings of information strategy (£)	2891	387	0	0	0	0	0	0	0	0	1945	2570
Year 3 2005–6	Infrastructure strategy (m^3)	6570	3350	0	0	0	0	0	0	0	0	4420	5840
	Information strategy (m^3)	0	2700	0	0	0	0	0	0	0	0	0	0
	Marginal cost excess of AWS supply over winter storage supply (£/m^3)	0.44	0.44	0.44	0.44	0.44	0.44	0.44	0.44	0.44	0.44	0.44	0.44
	Cost savings of information strategy (£)	2891	286	0	0	0	0	0	0	0	0	1945	2570
Year 4 2006–7 to 2022–23	As for 2005–6												

Table 4.7 Silver Birches plc: drought warning chart 2003–4.

Row		September	October	November	December	January	February	March	April	May	June	July	August
4	Required minimum combined flow Ely			319	319	319	319	114					
5	Ouse and Denver Sluices ('000 m^3)												
6	Required minimum combined flow			351	351	351	351	125					
7	Ely Ouse and Denver Sluices + 10%												
8	('000 m^3)												
9	Actual combined flow Ely Ouse plus	tbi	tbi	tbi	tbi	tbi	tbi	tbi	tbi	tbi	tbi	tbi	tbi
10	Denver Sluices ('000 m^3)												
11	1996 rainfall (mm)	391	391	391	391	391	391	391	391	391	391	391	391
12	1996 rainfall + 10% (mm)	430	430	430	430	430	430	430	430	430	430	430	430
13	Actual rainfall during last 12 months	tbi	tbi	tbi	tbi	tbi	tbi	tbi	tbi	tbi	tbi	tbi	tbi
14	inclusive of current month (mm)												

Note: rainfall data should be for the whole of the Anglian Region. tbi: to be inserted.

Next a drought warning chart was prepared for Silver Birches plc and this is set out in Table 4.7. In respect of pumping to the winter storage reservoir, rows 5 and 6 are the most important. If the actual values of row 6 prior to the 'pumping season' (November–March) fall short of 351,000 m^3 (10% higher than the minimum required for pumping to take place), management action is necessary. The consultant recommended that the rate of pumping to storage should be as early and as high as permitted and practical from 1 November onwards in order to secure the 36,000 m^3 quantity permitted under the licence. If on 1 November the actual combined flow is below the Environment Agency's required minimum of 319,000 m^3, then the farm would have to await the time when pumping *is* permitted. In the meantime in any months between November and March when the Agency prohibits abstraction to the reservoir, the mains supply must carry the maximum burden of 520 m^3 irrigation over those 5 months (see Table 4.5 for the year 2005–6).

In respect of rainfall, rows 11 and 12 are the most important. Row 12 (like row 6) is intended as a trigger mechanism for the farm's adaptive management. When actual rainfall falls short of 430 mm, it warns management that the Anglian Region's rainfall in the preceding 12 months has been only 10% in excess of the lowest level in the past 40 years. The management team should then decide whether to stay with the conjunctive supply balance of Table 4.5 or whether to prepare for that most unlikely event – a mains cut-off. If the latter decision is made, then the team should specify at what level of actual precipitation to take action and for what maximum length of duration of the cut-off they should prepare. Let us suppose that if the row 12 datum sinks below 411 mm (1996 + 5%), Silver Birches plc management takes action to guard against a 2-month mains cut-off. What it must then do is immediately to purchase sufficient water from AWS add to the existing stock of water in the winter storage reservoir so that there is sufficient water in storage to cover the entire irrigation requirements of the next 2 months. In this case, provided that any cut-off is indeed less than 2 months, the irrigation supply will remain unaffected by the drought. In that year, the annual advantage in AWS water purchased of the informational strategy over the infrastructural strategy will be reduced. Between 1963 and 2002 there were only two calendar years when rainfall fell below 411 mm; these were 1972 (409 mm) and 1996 itself (391 mm).

4.6 CONCLUSION

The purpose of this paper has been to show that drought management is not only a responsibility of public-sector water policy but can also be carried through by agribusiness itself with respect to net irrigation requirements. Farm-level planning of this kind becomes more important as climate change brings with it a greater proportion of extreme events and more frequent low dry season flows.

The case-study is set in the Anglian Region of the England and Wales Environment Agency, where 58% of the two countries' most productive agricultural land is found. This region is also the driest in the UK. The farm that is the focus of the paper is the largest producer of container trees in Europe. The growth of its trees in their containers is currently almost entirely dependent on irrigation water sourced from the mains supply of AWS, the private regional water utility.

Silver Birches plc has some 94,000 trees. An interruption of the mains supply during a drought would irreparably damage the entire stock within 15 days. Although such a cut-off is unlikely, the cost to the farm would be so great – immediate financial ruin – that drought management planning is now under way. The core element is to introduce conjunctive supply based on a farm-level winter storage reservoir. Moreover such water promises to be much cheaper in prime cost terms than the mains supply. Farm management's initial intention was to set aside 50% of the reservoir's capacity on a permanent basis so as to provide a fall-back in time of drought. This was called the 'infrastructural strategy'. A simple model was developed to show the monthly pattern of conjunctive supplies in this case.

An alternative 'informational strategy' was next developed using local and regional time-series data on surface water flows and precipitation. Based on these two variables, a drought warning chart was drawn up that allows farm management in most years to take full advantage of its winter storage reservoir. In drought periods it stimulates early action to protect the reservoir's capacity as a defence against a mains cut-off. The use of adaptive management triggers in this chart has, for the economist, striking parallels with the crisis signals used in models of international currency trading (Kaminsky et al. 1998; Dow 2002).

Comparing the two strategies, the use of a drought warning chart offers considerable savings in expenditure on mains water. Silver Birches plc is now setting up a small, on-farm weather station to underpin the informational strategy as well as to strengthen management understanding of crop water requirements for optimal irrigation scheduling.

The conclusion of this paper is that farm-level drought management should draw not only on civil engineering initiatives but also on an informed understanding of the often complex linkage of regional hydrology to the farm's hydrosocial supplies. Informational strategies, combined with infrastructural investment, can be cost-effective and by the intelligent use of adaptive management based on trigger routines can reduce or even eliminate the impact of drought on farm output.

5

The potential role for economic instruments in drought management

'And thus the whirligig of time brings in his revenges.'

William Shakespeare, *Twelfth Night*

5.1 INTRODUCTION

It is now widely accepted that human society generates greenhouse gases on such a substantial and increasing scale that climate change and global warming in the future will be of the greatest importance in our lives as well as in the lives of succeeding generations (Watson et al. 2001). Climate change, it is believed, will have marked impacts on water stocks and flows and this in turn will require innovations in the management of coastal and estuarine defence, drought, flooding, irrigated farming, nature conservation, storm and waste water systems, and water resources (Willows & Connell 2003: Table 3.1). In particular, the mean and variance of climate variables in the future may be associated with shorter return periods of climate extremes than in the past, for example of frost days or heavy rainfall events.

The focus of this paper is on drought management and it recognizes the risk that the return frequency of drought may rise during the 21st century (van Hofwegen & Svendsen 2000: 9–10). Drought is here defined as a sustained and regionally extensive occurrence of below average precipitation. The article's focus is sharpened by narrowing attention primarily to the *economic instruments* of drought management within the context of the economic impacts of drought events.

5.2 THE WATER ECONOMY

Water is essential to all forms of life, of course, as well as to all the activities of human society. But if we limit ourselves only to the part it plays within a specified *economy*, we can distinguish just three broad sectors. First, there are the rainfed areas of farming, forestry and pastoralism. Secondly, there is the instream sector with its navigation, fishing, conservation, recreation, tourism services and hydroelectric power production. Thirdly, we have the outstream sector where water meets the needs not only of households but also those of agriculture, mining, manufacturing, construction, public and private services.

In this paper the author will use the term 'the water economy' to refer to a region or country's economy, i.e. its production of goods and services, from the point of view of the dependence of output on water stocks and flows in the rainfed, instream and outstream sectors. The article in fact focuses only on the outstream sector. The impact of a drought on the outstream water economy can be twofold. It may be that there is an increase in the economic demand for water. This is true in the case of the irrigation sector, for example, as well as households' use of hosepipes and sprinklers. However, drought may simultaneously lead to restrictions on water supplies from the local utility with sharp negative impacts on companies' sales and on households' welfare.

So the orientation of what follows is to explore:

1. Drought management's degree of understanding of the impacts of drought on the outstream water economy.
2. How drought managers might in the future deploy the economic instrument of water charges to achieve their objectives.

The paper illustrates its principal theses from England's Anglian Region.

5.3 THE ANGLIAN REGION

For environmental policy purposes, England and Wales has a public sector body known as the Environment Agency, with its head office in Bristol. The Agency operates through eight regions, one of which is Anglian. The Environment Agency Anglian Region (EAAR) stretches from the Humber Estuary in the north to the River Thames in the south, from the Norfolk coast in the east to Northampton in the west. It covers 27,000 km^2 and five million people live in the area. The region has extensive, sparsely populated rural areas, particularly in the north and east, a long coastline and interspersed urban centres around which industry has developed. As communication links continue to improve (especially with London) it is likely to be one of the highest population growth areas in the country.

The hydrology of the region is of considerable interest. Annual average rainfall is 595 mm compared with a national average for England and Wales of 897 mm. Effective rainfall is only a quarter of the total at 147 mm and long, dry summers during which evaporation exceeds rainfall are a normal part of the region's climate. Plotting the cumulative difference between rainfall and potential evapotranspiration, the Agency shows that in the 100 years beginning in 1899, there were 32 years exhibiting a rainfall deficit (EAAR 2003: Fig. 3.6).

Twenty per cent of the area lies below sea level, including the extensive area of the Fens with slow-flowing or ponded rivers and extensive washlands. The main aquifers are the chalk underlying large parts of the east, limestones in the west and sandstones in parts of the centre and east (ibid. 20–2).

The Anglian Region has 58% of the most productive agricultural land in England and Wales and agriculture has considerable influence on the rural economy and communities. The sustainable use of water is crucial given the competing demands for it, particularly because this region is the driest in the UK. Low flows in late summer and increased demand especially for agriculture and garden use are likely, says the Environment Agency, to cause stress (www.environment-agency.gov.uk/). In some summers, irrigation can make up 50% of total use.

Alongside the Environment Agency, the second principal actor in water resources management in the region is Anglian Water Services Limited (AWS), which serves the needs of over five million industrial, commercial and domestic consumers (www.anglianwater.co.uk/). AWS is the main regional utility and its principal activities are the provision of drinking water and the treatment of waste water. The company employs 3700 full-time equivalent persons and in 2002 had a turnover of £724 million (AWS 2002). Its principal infrastructural assets are mains and sewers, impounding and pumped raw water storage reservoirs, dams, sludge pipelines and sea outfalls. The company's capital investment plan in the 5 years through 2004/5 is equal to £1.5 billion and covers further improvements to drinking water, bathing and river water quality, as well as network improvements to maintain serviceability and meet new demands.

In addition to AWS there are four other small water utilities in the region. They are Cambridge Water Company, Essex and Suffolk Water, Tendring Hundred Water Services, and Three Valleys Water. Overall, for these five utilities, 40% of public water supplies are provided from groundwater, mainly from the chalk and limestone aquifers, and 60% from surface water.

5.4 THE REGION'S WATER ECONOMY

The water economy of the region with respect to its outstream flows is best understood by examining its principal components.

5.4.1 Households

Anglian region has a population of some five millions persons. At the end of the 1990s, public water supply leakage and losses were one-fifth of the total public supply. With respect to use net of these losses, households took two-thirds of the total. Private domestic abstraction also exists, particularly in rural Norfolk. There is a significant difference in estimated use in litres per head per day (lhd) between metered and unmetered customers. In 1997–98 the data were 145 and 163 lhd, respectively (EAAR 2001: Fig. 5.3). National data for household use show the rounded breakdown between microcomponents as personal washing (33%), toilet (25%), clothes washing (14%), miscellaneous (13%), dish washing (8%), garden watering (6%), and car washing (1%) (ibid. Fig. A3.1).

Table 5.1 Anglian region irrigated agriculture.

Crop type	Water use (%)	Area irrigated (%)
Potatoes	57	41
Field vegetables	17	20
Sugar beet	15	22
Cereals	5	11
Fruit	3	3
Grass	3	3
Total	100	100

Source: EAAR 2001, Figure 3.13.

5.4.2 Agriculture

This sector ranks second in Anglian water use, accounting for about 10% of the total. Farmers use water for general purposes, livestock watering (particularly in pig rearing), and for irrigation – the largest consumer. The proportion of water used that is recycled back after use to the Anglian catchments is low in comparison with all other users. The supply sources are both mains provision as well as direct (surface and groundwater) abstraction. An interesting feature of the region is the operation of the Internal Drainage Boards. In the winter, water is pumped out of the drains to maintain the fields in a workable condition. In the summer water is diverted from the rivers to the drains whence it is distributed to the fields to maintain high water tables for irrigation pumping (EAAR 2001: 26, 49; Merrett 2002a: 105–12). Spray irrigation is the principal technology. Water is required mainly in the summer months when plant growth is most rapid and when transpiration cooling is critical. This is precisely the season when effective rainfall is at its lowest, as we have seen. Abstraction has to be even higher in drought years when the farmer loses his rainfed source. Table 5.1 shows the irrigated sector's crops by area and irrigation water use. Potatoes, field vegetables and sugar beet predominate.

5.4.3 Other sectors

Table 5.2 shows the 'other sectors' use of water. The volumes supplied in 1998 were about the same from the region's water utilities as from direct abstraction by users. Today the ratio is closer to two-to-one. Retail services, the production of food and drink, the machinery sector, chemicals, and the extractive sector are the main players. The public water supply is of drinking water standard and more costly than direct abstraction sources. Chalk groundwater is also favoured by industry when it seeks a high-quality supply. For cooling or lower quality washing requirements, river water is acceptable. The extraction industry's dewatering activities are mostly exempt from abstraction licensing; they also engage in direct abstraction for the washing of sands and gravels. In the case of these 'other sectors', as with households, their consumption losses are low.

Table 5.2 Other sectors' use of water in 1998 in the Anglian Region.

Sector	Public water supply (%)	Direct abstraction (%)
Retail	32	0
Food and drink	21	34
Machinery	16	1
Chemicals	8	19
Education and health	8	0
Hotels	7	0
Other*	5	4
Transport	3	0
Extractive	0	38
Construction	0	4
Total	100	100

* Includes paper, metals, textiles and utilities. Source: EAAR 2001: Figure 3.12.

5.5 THE AGENCY'S DROUGHT PLAN

The Environment Agency's Anglian Region staff are, of course, familiar with the risk of drought in their area. Droughts may take the form of hot, dry summers or of a longer period of low rainfall and dry winters, as in 1989–92. Extended dry periods affect rivers quickly but flows are maintained by groundwater baseflow and wetlands. As these sources are exhausted river flows drop to very low levels, their temperature increases, there is less dilution of effluent and oxygen levels fall. Fish spawning grounds may be damaged and water supplies are threatened at the same time as irrigation needs are increasing greatly. The return of rainfall does not end the low flows for the dry soil soaks up the rain and many weeks may pass before there is a sustained rise in river and groundwater levels.

The Agency has pointed out the strong relation between climate change and a regional drought.

> Changes in climate will ... change groundwater and river flow regimes and therefore the availability of water for abstraction. Current estimates of climate change suggest that by the 2020s throughout the Anglian Region there will be, on average, more winter rainfall and less summer rainfall. Effectively, this means that the climate will be less predictable, with both more dry years and more wet years. This in turn means that low flows will probably occur more often. Evidence of the possibility of longer droughts is unclear; the best available view appears to be that increased variability makes drought that lasts over several years slightly less likely. However, it is important to note that the understanding of changes in extreme events is more limited than that of changes in average climate. (Environment Agency Anglian Region 2001: 47)

The Environment Agency Anglian Region published its drought plan as recently as May 2003 (EAAR 2003). Drought plans are required as part of the Agency's statutory duty to secure the proper use of water resources. The plan's chapter heads are: (i) drought management teams, (ii) drought monitoring arrangements, (iii) actions and mitigation, (iv) corporate affairs, and (v) reporting. Careful reading of the text shows that the Agency's principal policy instruments in drought management are informational, infrastructural and regulatory. These include, for example, public relations media releases, regional transfer schemes such as the Ely Ouse-Essex Scheme, and abstraction limitations.

There are no economic instruments for drought management of any kind. Indeed the Plan specifically states (ibid. 14):

> Abstraction charges are set at a higher rate than would be merited by moderate to wet conditions to ensure surplus money is available for financing the operation of [transfer schemes] during dry years. This avoids large increases in rates that would be required when a drought occurs.

With respect to this paper's interest in assessment of the water economy, there are many indications of the Agency's concern for specific sectors. This builds on the Region's excellent 2001 *Water Resources for the Future: a Strategy for Anglian Region*. There is an evident wish to provide timely warnings, particularly for spray irrigators in relation to the likely need for implementation of restrictions on the use of their licences under Section 57 of the Water Resources Act 1991. A leaflet 'Prospects for Spray Irrigation' sets out the communication channels and procedures for introducing irrigation restrictions. The importance of communications between Agency staff and the farming community is highlighted and, when drought occurs, local meetings will allow the Agency to discuss directly with abstractors how best to instigate spray irrigation restrictions to minimize impact on the farming community (EAAR 2003: 27, 41). Restrictions from surface water often take the form of each licensee being able to use his/her daily quantity on *alternate* days. Night-time irrigation is encouraged. 'The Agency also expects those irrigators with summer licences and a winter storage reservoir to stop summer abstractions and use their stored winter water'. Moreover, trialling of a new cost-benefit method for assessing spray irrigation restrictions has been conducted on susceptible sites across the region.

Water companies such as Anglian Water Services are also permitted to seek authorization to increase their rate of abstraction prior to a likely drought event. The Agency also states that it '... is committed to encouraging Water Companies to adopt measures for demand management and efficient use of water' (ibid. 24).

The drought plan makes only the briefest reference to other groups of abstractors. However, it is pointed out that many have been previously targeted to be more efficient with their water use. Local authority Environmental Health Officers contact households with non-licensable domestic sources in rural areas where the possibility of dry wells exists. Overall the Agency sees itself as 'balancing demand for a limited supply from various quarters, while being viewed publicly as arbitrators on behalf of the environment' (ibid. 45).

5.6 ANGLIAN WATER SERVICES' DROUGHT PLAN

AWS published its drought plan in March 2003 (AWS 2003). AWS is required by the UK government's Department of Environment, Food and Rural Affairs (DEFRA) to agree updated drought plans with the Environment Agency. Water company drought plans have become a statutory requirement in the Water Bill 2003. The Plan's section heads are: (i) introduction, (ii) background to previous droughts, (iii) water resources planning for a drought, (iv) drought management (surface water and groundwater), (v) demand-side management, (vi) supply-side management, (vii) mitigation of environmental impact and (viii) summary. Careful reading of the text shows that the AWS's principal policy instruments in drought management, *just as with the Agency*, are informational, infrastructural and regulatory. These include, for example, publicity campaigns, the construction of pumped, surface water storage reservoirs with long retention periods, and hosepipe bans.

There are no economic instruments for drought management of any kind. This is in spite of the fact that AWS is clearly convinced of the general effectiveness of the management of economic demand. Between 1989 and 2003 the population of the AWS area increased by 400,000 persons, and 20,000 new properties have been connected each year. However the total quantity put into supply for customers has shown no upward (or downward) trend. In part this is due to the extension of household metering; 50% of AWS households now pay on the basis of volume used.

With respect to this paper's interest in assessment of the water economy, it is remarkable that (unlike the Agency) the AWS drought plan gives no indication of its concern for specific sectors. This is true even though the plan briefly refers to the possibility of 'the introduction of demand restrictions on surface water supplies in the event that reservoirs are drawn down to control curves' (ibid. 7).

In the preparation of its drought plan, the AWS was required to use the Drought Plan Guideline published at the national level by the Environment Agency (EA 2002). This Guideline in fact makes no reference to economic instruments of drought management nor does it suggest that utilities should review the economic impact on their customers of regulatory action.

5.7 DROUGHT PLANS AND THE WATER ECONOMY

Two principal institutions lie at the heart of this analysis, the Environment Agency Anglian Region and AWS. The Agency is the statutory body with a duty for strategic water resources planning. Its role is to protect the long-term future of the water environment while encouraging sustainable development. Its vision for water resources in the next 25 years is abstraction of water that is environmentally and economically sustainable, providing the right amount of water for people, agriculture, commerce and industry, and an improved water-related environment (EAAR 2001: 13). In contrast, AWS is a private limited company that supplies water and waste water services to its customers within the area of its operation and thereby earns profits for its shareholders.

Despite this immense difference in roles between the two institutions, we can see certain characteristics common to their separate drought plans. In the first place we have a clear commitment by each to cooperate with the other in drought management. Secondly, in managing a drought event both institutions deploy policy instruments that are informational, infrastructural and regulatory. Thirdly, neither institution deploys an economic instrument in the management process. Fourthly, both institutions discriminate between user types in reducing the volume of water used. The Agency may permit AWS to increase its abstractions prior to a likely drought event; and spray irrigators may encounter restrictions on the use of their licences. AWS may institute hosepipe and sprinkler bans (at the request of the Agency) as well as bans on 'non-essential use' such as 'vehicle washing, irrigation of ornamental gardens, etc.' (EAAR 2003: 24). Finally, the drought plans of neither institution seek to demonstrate that their selection of specific actors for regulatory restriction is demonstrably the optimal solution or a satisficing arrangement that, in some sense, minimizes the aggregate welfare loss of such restrictions.

5.8 ECONOMIC INSTRUMENTS

In this section I propose that the three principal national institutions for water resources management should review and evaluate the feasibility, costs, benefits, risks and uncertainties of replacing the regulatory instruments of drought management by economic instruments. These three institutions are DEFRA, the Environment Agency and the Office of Water Services (Ofwat).

The final sections of this paper will provide a preliminary overview of six key issues that are likely to arise in the review/evaluation for which I call.

5.8.1 Legislation

In this context 'economic instrument' refers to any one of three ways in which the EAAR or AWS (as well as the four other regional water utilities) require outstream water users to pay for their water supplies. These forms of payment are:

- A charge for abstraction, based on the metered volume withdrawn.
- A price per cubic metre of metered mains water supplied to users.
- A charge for unmetered mains water supplied to a household's house or flat, where the charge is based on the capital value of their dwelling.

The proposal here is that, in comparison with the non-drought situation, these three forms of payment, which are already practically universal in England and Wales, *would be successively raised at what the Environment Agency calls the potential drought stage, then again at the established drought stage and finally again at the severe drought stage* (EAAR 2003: Table 1). The charge increases would be removed as soon as the post-drought wind-down is declared. Authority for the Environment Agency and the water utilities to take such action would require fresh legislation in the British parliament as well as new powers for the Office of Water Services in Birmingham.

5.8.2 Technical issues: time

The charges of types 1 and 2 above are based on metered flows. They can be introduced only where it is technically feasible to measure the volume of water supplied to users during each of the three drought phases referred to above (potential, established, severe). This requires meters capable of providing the volumetric flow on each day of the month. A variety of such 'real-time' meters are available on the market. For charge-type 3 above, which at present applies to 50% of Anglian Region households, the additional charge would be based on the number of days that fell within the three drought stages during the billing period. It has to be said that the economic management of demand is always more difficult in the absence of metered supplies (Merrett 2002b). Therefore, with an unmetered supply, there exists no economic incentive for the household to moderate the quantity used during a drought, even though the daily charge is raised.

5.8.3 Technical issues: place

In England and Wales the location of each abstraction point for surface water or groundwater is precisely known. It is also true that a feature of droughts is that they vary in severity and environmental impact across any given region as well as varying in severity and impact with source type. For example, for water resource management purposes the Agency divides the Anglian Region into 183 surface water catchments and 64 groundwater units (EAAR 2003: 15–17). It would be both possible and desirable, therefore, to vary the increase in charge-type 1 above with the location of the point of abstraction.

5.8.4 Economic: rationale

The basic principle is that the regulatory instrument of rationing specific user-types (such as spray irrigators or households with gardens) be abolished and replaced by the economic instrument. Environment Agency and water company decisions on who should bear the brunt of supply shortfalls during a drought would be replaced by an instrument whereby those who are most willing and able to pay for the reduced volume supplied would receive it. Regulatory fiat would be replaced by users' decisions. However, it is essential that households do not face a price or charge increase for their 'lifeline' supply.

5.8.5 Economic: demand response

Deployment of the economic instrument would permit the EAAR and the AWS to vary the water prices and charges it sets throughout the drought's duration and across the region's surface water catchments and groundwater units. The change in volume of water purchased following a change in price of the metered supply is described by economists as the elasticity of demand (Merrett 2002a: 23–5). Both the Agency and the utilities would wish to carry out new research into the economic demand for water in the region by different users so as to be able to estimate the price rises they would apply at each stage of the drought. The main actors should be able to vary their volumetric prices and charges within a high upper bound set jointly by the Environment Agency and Ofwat. Morris and his colleagues at Cranfield University's Institute of Water and Environment have recently shown that in England's irrigated agriculture 'although increased water charges can reduce demand in low value applications, very large increases in charges are needed to reduce water demand where the financial returns to water are high, resulting in large decreases in farm incomes ... before consumption significantly changes' (Morris et al. 2003: 623). Meanwhile, Chapter 4 in this book shows how economically devastating regulatory action would be for a specific Anglian farm during a drought, as well as the climate adaptation options that farm management can pursue.

5.8.6 Political

Is this proposal likely to be acceptable to the people of England and Wales? I believe that the shift of power from regulator to user will be welcome. However, two issues of particular importance would have to be addressed. *First*, the price paid by households for their lifeline supply should be unchanged, for social reasons. *Secondly*, what is a reasonable but effective way to handle unmetered users? The importance of this fades the more successful is national government in increasing the proportion of households that *are* metered. As David King, Director of Water Management of the Environment Agency, writes in response to the first draft of this paper: 'You are right in thinking that at present drought management makes little use of economic incentives to control demands. For households far more meter penetration would be necessary before this became a viable option. However, we will keep this under review and we will consider all options that can help with effective water resources management' (pers. com. 3/12/2003). George Day, Head of Supply/Demand Balance at Ofwat expresses a similar opinion in a letter to me.

5.9 CONCLUSIONS

Climate change will likely be associated in the future with shorter return periods of climate extremes, including drought. So effective drought management is set to become even more necessary in the future than it is already. This in turn demands a review of the current instruments of policy that we deploy as well as a better understanding of their impact, not only on the rainfed and instream sectors, but also on the outstream water economy. The argument of this paper is developed for England's Anglian region, particularly with respect to the roles of the Environment Agency and of AWS.

Review of these two institutions' drought plans shows that their instruments of policy are informational, infrastructural and regulatory. There are no economic instruments of any kind. Moreover, the Environment Agency at the national level in its drought plan guideline makes reference neither to utilities deploying economic instruments in demand management nor to a need for utilities to review the impact on their customers of regulatory action.

Yet both the Environment Agency at the regional level as well as the AWS discriminate between user types in reducing the volume of water used. But neither institution seeks to demonstrate that its selection of specific actors for regulatory restriction is demonstrably the optimal solution or a satisficing arrangement that, in some sense, minimizes the aggregate welfare loss of such restrictions.

With respect to outstream water, the author concludes that the three principal national institutions for water resources management, DEFRA, the Environment Agency and Ofwat, should review and evaluate the feasibility, costs, benefits, risks and uncertainties of replacing the regulatory instruments of drought management by economic instruments. If that review and evaluation opts for economic instruments, they would shape both the location of abstraction as well as the relative shares between outstream water users, based on the responsiveness to charge and price changes for the abstraction and use of water.

The material in this chapter originally appeared in: The potential role for economic instruments in drought management. *Irrigation and Drainage*, 53: 1–9, 2004.

6

'Virtual water' and Occam's razor

Until then I had thought each book spoke of the things, human or divine, that lie outside books. Now I realized that not infrequently books speak of books: it is as if they spoke among themselves.

Umberto Eco, *The Name of the Rose*

6.1 INTRODUCTION

In 2001 Professor Tony Allan of the University of London's School of Oriental and African Studies saw published his *magnum opus* entitled *The Middle East Water Question: Hydropolitics and the Global Economy*. The book marks the final step in his long march from geography to the politics of water resources. It also provides the most complete statement of his views on the rôle of virtual water in linking regions of water scarcity with regions rich in water. In fact, his main interest is the Middle East and North Africa (MENA). The concept of 'virtual water', virtually synonymous with Allan's name, has been taken up increasingly widely in recent years and the time is ripe for a critical review of its relevance to our understanding of the manner in which the water needs of semi-arid countries are met. Such a review is the object of this paper. Section 6.2 provides a résumé of the main propositions of the virtual water thesis. Section 6.3 deploys the main critique. Section 6.4 introduces Occam's razor. Section 6.5 presents my conclusion.

Before getting under way it is worth noting that the disciplinary foundation of this paper is philosophy. So a useful philosophical convention will be used. Whenever I am discussing a *concept*, the word will be placed in single inverted commas. For example: "Spinoza tells us that the concept 'dog' cannot bark". Whenever I refer to the *thing* or *activity* to which a concept such as 'dog' refers, there are no inverted commas. For example: "Hagrid's dog Fluffy barked without ceasing".

 The Price of Water: Studies in Water Resource Economics and Management by Stephen Merrett. Published by IWA Publishing, London, UK. ISBN: 1 84339 081 7.

Two other features of a philosophical paper are worth noting. *First*, philosophers within the English tradition are primarily concerned with the relation between language and truth, so they commonly write in a dense and compact style where the text needs to be read at a measured pace to be understood. *Second*, non-philosophers may take the view that philosophy is merely rhetorical quibbling, nothing but words, words, words. My view is different. Water resources management has traditionally been founded on engineering and hydrology – with a history of immense achievements. But we now recognize that social and environmental scientists should ally themselves with the hydrologist and the engineer in the challenges that the future holds for us all. This requires that political scientists and economists, for example, develop an appropriate language for their work. They are now in the stage where such a language is still in the furnace. The role of philosophy is to assist in the forging of this new language, in which the objects of criticism may be water 'efficiency' in irrigation or water 'demand' in urban planning or 'virtual water' in global trade or the 'contingent value' of a buttercup. In a nutshell, the successful advance of science is critically dependent on progress in its language.

6.2 A WATER DEFICIT RESOLVED

A brief, neutral statement of the virtual water thesis is set out in this section, beginning with Allan's concept of a region's 'water deficit'. 'Region' here refers to any relevant geographic area including a country, a province or a catchment.

For any region it is possible to estimate whether or not it suffers from a water deficit in the sense that 'there is not enough surface water, ground water and soil water to meet the domestic, industrial, municipal and food needs of its population' (T. Allan pers. comm.). In the specific case of what will be referred to as Region A, its surface water, groundwater and soil water are insufficient to meet the needs described above. This shortfall of supply with respect to need is Region A's water deficit (Allan 2001: 30).

In fact the needs of households and industry can and are comfortably met by regional supply. It is the crop water requirements of food self-sufficiency that are impossible to satisfy. As a consequence the water deficit is resolved by the import of virtual water from Region B. 'Virtual water' refers to the soil water sourced by precipitation and irrigation that meets the crop water requirements of the food grown in Region B that is subsequently exported to Region A. Region A's imports of such food are sufficient, alongside domestic production, to meet entirely the food needs of its population.

To indicate the scale of these virtual water imports, we can take the example of wheat. One tonne of exported wheat requires about one thousand tonnes of virtual water (Allan 2001: 106, 126). Less than 0.1% of the virtual water is physically embedded in the food grains themselves. During cultivation in Region B, more than 99.9% of the virtual water returns to the irrigation cycle as farmland drainage or is lost in evapotranspiration. As Allan writes (2001: 106):

> At the 1000 tonnes (cubic metres) of water per tonne of grain estimate of water content the [MENA] regional imports of virtual water by the mid-1980s were equivalent to the annual flow of the Nile into the Egyptian agricultural sector.

6.3 A CRITIQUE OF THE VIRTUAL WATER THESIS

Allan's thesis summarized in section two is, I suggest, flawed in a number of ways. The critique will be set out with respect to the use of the term 'virtual water', the import of virtual water and, finally, the farm sectors of Regions A and B.

Virtual water. In everyday English we use the word 'virtual' (as in 'virtual reality') to mean something parallel to or imitative of a real-life entity or process. But Allan's 'virtual water' does not do this. This is because virtual water is real water; it is the soil water of Region B used in meeting the crop water requirements of that Region's food exports. This neglect of a well-established linguistic routine has the result that the central concept of the argument, 'virtual water', is not virtual in any sense. It is not good science to build theory on terms that are inherently misleading.

6.3.1 The import of virtual water

In *The Middle East Water Question: Hydropolitics and the Global Economy* we repeatedly read that virtual water is imported into the MENA region. So one expects to see each single ship carrying a full cargo of wheat, maize or rice imported from the wide world and heading for the sunlit harbours of the Mediterranean and the Red Sea to be followed by one thousand additional ships each carrying in its hold a full cargo of the virtual water used for the imports' crop water requirements. A magnificent sight – but one we are unlikely to witness in our time. It is as if, having identified a water deficit in Region A, virtual water must be imported to eliminate it. It may be that Allan does not wish us to believe literally that this happens. Clearly propositions of the type "water security for the MENA region is achieved by virtual water imports" are false, and as misleading as the tabulation of 40 billion cubic metres of cost-free water transported into the region each year (Allan 2001: Table 2.1). But if such statements are, indeed, merely metaphorical, they cannot at the same time be read as part of a scientific argument on international trade in food and water.

6.3.2 Farming in Regions A and B

When one approaches agriculture from the perspective of water resources there is a danger that the experience of farming is seen largely with respect to its crop water and net irrigation requirements. As a result, a more rounded vision is lacking, one that understands that the water theme is only one amongst many, such as soil characteristics, land rights, labour skills, pest control, farm budgets and product markets. Consequently, if we use the term 'the import of food', this opens up major questions rendered invisible by 'the import of virtual water'. Have food imports led to higher population birth rates in water deficit regions than would have occurred in their absence? Do food imports weaken the farm sector of the importing country? Do food imports open the importing country to political control from the exporting country? Will the importing country be able to maintain its foreign exchange expenditure on food imports in the long run? Finally, if food imports are subsidized when they are shipped from the European Union and North America, will these subsidies be maintained in the long term?

6.4 OCCAM'S RAZOR

So we come to William of Occam, a Franciscan scholar and Aristotelian philosopher who lived between *circa* 1295 and 1349, the year of the Black Death. He took part forcefully in the great debates of his time on the poverty of the clergy and on transubstantiation and was excommunicated by Pope John XXII in 1328. Bertrand Russell writes of him:

> Occam is best known for a maxim which is not to be found in his works, but has acquired the name of 'Occam's razor'. This maxim says: 'Entities are not to be multiplied without necessity.' Although he did not say this, he said something which has much the same effect, namely: 'It is vain to do with more what can be done with fewer.' That is to say, if everything in some science can be interpreted without assuming this or that hypothetical entity, there is no ground for assuming it. I have myself found this a most fruitful principle in logical analysis. (Russell 1996: 462–3)

I now wish to explore whether Occam's razor can properly be applied in excising two of Allan's core concepts: 'virtual water' and 'the import of virtual water'.

'*Virtual water*'. This term can be replaced by 'the crop water requirements of food exports'.

'*The import of virtual water*' can be replaced by 'the import of food'.

More generally, the huge economic, political and social processes that are addressed by Allan in the terminology already described can be reset in a world where semi-arid (and other) regions do not have the capacity to feed their populations and so import food. These imports mean that less production and therefore less water is required in these regions' irrigated agriculture. Regional politicians may deflect attention from such dependence; the availability of imported food allows them both to postpone new water supply initiatives and to delay difficult decisions about the demand management of their water resources.

6.5 CONCLUSION

In this paper it is argued that:

- 'Virtual water' refers to real water – there is nothing virtual about it. It denotes the crop water requirements of food exports.
- 'The import of virtual water' is a metaphorical term, not a scientific one, and its use leads to statements that are plainly false. It denotes the import of food.
- In its policy applications, 'the import of virtual water' leads to a neglect of the current and future status of the agricultural sectors of the countries importing and exporting food.

My conclusion is that water resource researchers and policy-makers should apply Occam's razor to the virtual water thesis.

The material in this chapter originally appeared in: Virtual water and Occam's razor. *Water International*, 28(1): 103–105, 2003.

7

Virtual water and the Kyoto consensus

A man may imagine things that are false, but he can only understand things that are true.

Isaac Newton, *Theological Manuscripts*

7.1 THE USE OF METAPHOR

Three papers were published in the March 2003 edition of *Water International* that sought to clarify and evaluate the use of the concept 'virtual water' in our understanding of water resources management in water-deficit catchments, regions and countries (Allan 2003, Lant 2003, Merrett 2003). In the following discussion note this author seeks to capture the main points of his continuing differences with Tony Allan.

Allan and I are in complete agreement that phrases such as 'the import of virtual water' are metaphors, not propositions that can be simply said to be true or false. Moreover, we now have a clear definition to work with. 'Virtual water is the water needed to produce agricultural commodities' (Allan 2003: 107). It follows that Allan accepts that there is absolutely nothing 'virtual' about virtual water. Virtual water is real water. It is, first, the water needs of livestock. Second, it is the soil water required to grow crops. Soil water is sourced by rainfall, irrigation practices and seasonal flooding such as that of the River Nile (Allan 1995: Table 1, Merrett 2002: Table 3.1).

In the original paper I suggested that water resources management is increasingly becoming a field in which social scientists play a part, alongside our colleagues from engineering and hydrology for example. The language we use to understand that most simple, yet complex of entities – water – is still being forged; and the successful advance of science is critically dependent on progress in its language.

Here we come to the crux of the argument. I believe of course that there is a place for metaphor and other linguistic tricks in published papers and books as a means to entertain, lighten the text and communicate in a telling manner. This paper itself contains a number of metaphors. The question is, do we advance our capacity to understand water resources by introducing a new term when we already have a perfectly good one? If we can speak of the 'water requirements' of agricultural commodities, why do we need 'virtual water'? Moreover, can it be wise to base our analysis on a single, powerful metaphor – that virtual water is imported?

7.2 CROPS, CROP WATER AND WATER DEFICITS

But there is a second arrow to my bow. 'Virtual water', as a central and constantly reiterated concept in the discussion of arid and semi-arid countries water requirements, can mislead us, create serious errors of analysis. I illustrated this in my March 2003 paper by pointing out that Allan in his book *The Middle East Water Question: Hydropolitics and the Global Economy* (2001: Table 2.1) includes in his 'typology of Middle East and North African waters' 40 billion m^3 per year of transported water in the form of virtual water, alongside the flows of pipelines, tankers and bags. His use of the 'virtual water' concept led directly to a 40 billion cubic metre error. This is because the 40 billion cubic metres to which he refers are the crop water requirements (primarily of foodgrains) sourced by rainfall and irrigation in the foodgrain-exporting areas such as Australia, Canada, the European Union and the USA. Clearly these are *not* 'Middle East and North African waters' and they are *not* 'transported' to the MENA region. Allan confuses crops with crop water; it is the former that are transported not the latter.

This inbuilt propensity to error appears in a new form in Allan's 2003 paper, as I shall show. We would all of us probably accept that there is some usefulness in the idea of a region or country suffering from a water deficit. This is defined by Allan, for example, as a situation in which there is not enough surface water, ground water and soil water to meet the domestic, industrial, municipal and food needs of its population (Merrett 2003: 104).

Many of us would say that one course of action, given such a present deficit, is to import food. For example, the London region has done this successfully for 2000 years. For Allan, this misses the point. In his 2003 paper he suggests that access to virtual water *remedies water deficits* and, with freshwater and desalinated water, *achieves water security*.

> … 10 percent or so of water needed for drinking, domestic, industrial, and service uses must come from freshwater or from manufactured sources. The 90 percent of water need for food and other agricultural production can come from freshwater, from soil water or it can be accessed in effect via food imports. Virtual water and manufactured water are the very successful means by which water deficit economies can remedy their deficits. (Allan 2003: 108)

Once again we see a dangerous crossover between talking about food imports as if they were water imports. In matter of fact, food imports tend to *increase* a region's water deficit. This is of great importance in semi-arid countries with dense populations. The growth in the water deficit arises from the faster rate of growth of population that imported food permits. In addition, lower levels of national production of food follow from the lowering of food prices brought about by imports. This is particularly true when imported food is subsidized and when the importing country's exchange rate is over-valued, making imports even cheaper.

In summary, I believe we should dispense entirely with the term 'virtual water'. This is for five reasons:

1. The term is redundant; virtual water is nothing more or less than the water needed to produce agricultural commodities.
2. There is nothing virtual about virtual water.
3. The use of the term leads in particular to a neglect of the impact of food imports on a country's agricultural sector, because food imports are falsely represented as water imports.
4. The confusion of water with food that accompanies the term 'virtual water' like a dark shadow fosters analytic errors such as that food grain purchases are transported water, that food imports remedy water deficits in the importing countries and that imported food brings 'total water self-sufficiency' for the importing region (Allan 2003: Figure 3). *The greatest weakness of Allan's concept is that it repeatedly confuses crop outputs with the water required to produce them.* If, as Allan suggests, Hoekstra & Hung (2002) conclude that 695 km^3 of virtual water is traded each year, why have none of us seen the boats within which it is carried? Why are they not causing gridlock in the world's sea channels? Virtual water is real water; if it is indeed traded it needs to be transported to the new owner. In fact, no such trade ever takes place in the case of food exports. Hoekstra and Hung, outstripping Allan, commit a 695 km^3 error.
5. Social science, if it is constructed on a metaphor, is built on sand.

7.3 THE KYOTO CONSENSUS

The final issue raised here is whether or not 'virtual water' has now entered the emerging consensus shaping current approaches to water resources management signalled in the World Water Forum in Kyoto, the 'Kyoto consensus' as I shall call it. Key features are that big dams are bad whereas small dams are good; that water resources management should be carried out on the catchment scale; that irrigation projects should be turned over to water user associations; that public water utilities are ripe for privatization; that water resources management should focus on demand management; and that water use should everywhere be priced. A thoroughly modern schema set, unfortunately, in a post-modernist world.

The Kyoto consensus is best understood as the waterish extension of the Washington consensus. What follows bears comparison with Christopher Lant's insightful paper, both his suggestion that 'virtual water' refers to a water resources management strategy and his comments on the economic linkage between the MENA region and the dominant grain-exporting regions (Lant 2003: 113–15).

The reasoning is this. The Washington consensus is that human society throughout the globe should, in its economic activities, be organized around free markets in which the dominant players are private companies with governments playing only a limited, regulatory role. Those free markets should include the export and import of goods and services between countries. The hegemonic political and economic power in all of this is to be wielded by the dominant classes of the USA and the European Union and their institutions, in particular the International Monetary Fund and the World Bank. Grain exports from Australia, Canada, Europe and the USA should be encouraged in this context, particularly to semi-arid countries with dense populations.

Two major problems arise from this specific agenda. *First*, as indicated earlier in this rejoinder, food imports by the South weaken the domestic sector that produces food by driving down the market price of grain output. The decline of agriculture drives down rural incomes and stimulates rural-urban migration. *Second*, where the South has the capacity to export services and manufactured goods on a substantial scale, food imports can be financed out of export earnings. But if the South cannot pay for its imported food in this way, it becomes dependent on North America and Europe for subsidy of these imported foodgrains. This means that, along with other forms of financial and military dependence, the future of these countries is shaped by the North. In this case such countries, far from looking to a future of total water self-sufficiency as the 'virtual water' theorists argue, face an indefinite journey of economic dependence. Here we find confirmation that the most important characteristic of economic systems is the complex forms that exist of the interdependence of supply and demand.

The material in this chapter originally appeared in: Virtual water and the Kyoto consensus. *Water International* 28(4): 540–542, 2003.

8

Catchment water deficits in the twenty-first century

Ye can discern the face of the sky; but can ye not discern the signs of the times?

St Matthew ch. 16, v. 3

8.1 INTRODUCTION

The objective of this paper is to provide a general theory of the manner in which the water resources of a catchment shift from surplus to deficit, and the means by which water resource institutions can manage or reverse this shift. The text requires the clarification of some key concepts and this is done as the argument proceeds. Note that 'catchment' and 'river basin' are used interchangeably. The term 'watershed' is avoided because it has rival and inconsistent definitions.

The broad *hydrological* framework here is perfectly conventional. Total precipitation in a catchment, where the principal river flows to the ocean, is consumed in part as evapotranspiration; the residual is referred to as effective precipitation. This latter flow recharges the basin's aquifers, the boundaries of which fall within the catchment. Effective precipitation also sources run-off, most strikingly in the case of the main river itself. The aquifers supply base-flow to the river system as well as discharging as springs. Finally, both surface water and groundwater discharge at the coast.

The broad *hydrosocial* framework is set out in Table 8.1. It represents only outstream flows. Outstream water is that which is abstracted by human society from lakes, rivers and aquifers, to be stored and channelled for human use. As Table 8.1 shows, both supply and use may have negative as well as positive entries. Total net supply is mathematically identical to total use; this is empirically confirmed only if all the supply and use entries are comprehensively assessed and accurately measured. The evapotranspiration of outstream (or instream) water is hereafter referred to as water consumption.

Table 8.1 A catchment's hydrosocial balance.

Row	Title	Base year volumes
4	**Categories of Supply (Positive)**	
5	Rainwater collection	a
6	Groundwater abstraction	b
7	Surface water abstraction	c
8	Desalination of sea water	d
9	Imports of water from other areas	e
10	Internal reuse of wastewater	f
11	External reuse of wastewater	g
12	Net fall in water abstracted and stored	h
13	**Total Supply**	**i**
14		
15	**Categories of Supply (Negative)**	
16	Supply-side evapotranspiration losses	j
17	Supply-side leakage	k
18	Exports of water to other regions	l
19	Net rise in water abstracted and stored	m
20	**Total Negative Values (Supply)**	**n**
21		
22	**Total Net Supply**	**i−n**
23		
24	**Categories of Use (Positive)**	
25	Households	o
26	Agriculture, including irrigation requirements	p
27	Mining	q
28	Manufacturing	r
29	Construction	s
30	Public services	t
31	Private services	u
32	Other uses	v
33	**Total Use**	**w**
34		
35	**Categories of Use (Negative)**	
36	Evaporation losses on users' properties	x
37	Leakage on users' properties	y
38	**Total Negative Values (Use)**	**z**
39		
40	**Total Net Use**	**w−z**

Source: Adapted from Merrett (2002a) Table 7.1.

It has already been stated that the paper addresses river basin surpluses and deficits. Therefore, these two concepts require definition. A catchment water surplus is a situation in which, throughout the course of a specified year, total precipitation in the basin is sufficient to simultaneously satisfy four conditions:

1. Abstraction from the aquifer is maintained at a sustainable rate.
2. Outstream water fully meets the economic demand for water from households, agriculture, mining, manufacturing, construction and the services sectors.

3. The basin population's economic demand for food is fully met from the domestic rainfed and irrigation sectors and/or from food imports financed by the basin's commodity and service exports and/or from domestic fisheries.
4. The river's instream flows do not fall below defined minima.

When a catchment water surplus does not exist, the river basin is in deficit. Whether or not conditions 1–4 are met is a matter of judgement. That is the human condition. For example, sustainable aquifer management is a multi-year activity because storage naturally fluctuates over time with annual effective precipitation.

The presentation of the general theory below uses a thought experiment applied to a fictional catchment under threat of a water deficit and continues by reviewing a wide range of redemptive options. In the light of what follows, the reader should recall that the *Oxford English Dictionary* cites J. Dwyer in 1847 as the first person to use the word 'catchment' (OUP 1979: 355).

8.2 THE DWYER CATCHMENT

Consider a river valley, with its many tributaries, that flows to the ocean. One can call it the Dwyer. Total precipitation, averaged over the years, has remained constant throughout living memory, despite climate change. The Dwyer catchment has been settled for centuries and contains a network of villages and towns as well as a single metropolis. For many decades the catchment population has grown at a steady rate such that total population increases by one-third every 25 years. Curiously, this rate of growth is equal to projected global population increase in the first 25 years of the twenty-first century (Merrett 2002a: 1).

The Dwyer River Basin is a source of considerable economic output through its hydroelectric power schemes, its navigation activities, its fishing catches and its location for conservation, recreation and environmental tourism. Moreover, diversions and pumping from the Dwyer and its tributaries, as well as spring capture and the pumping of groundwater have been sufficient throughout the basin's history to meet outstream use. The catchment is recognized currently to be in surplus.

The water law of 1847 protects the catchment's instream flows, its 'environmental needs' as some call them, in terms of both quantity and quality. The quantitative rule for the Dwyer's annual flow to the ocean is that it should never fall short of one-third of the past 10 years' average of net precipitation within the basin. This is referred to as the one-third rule. Net precipitation is the estimate of what effective precipitation *would* be, were there no dams, no hydrosocial supply and no hydrosocial use of water in the basin. This revision of *actual* effective precipitation is made in order to avoid prioritizing, for example, evaporation from reservoirs or irrigation water consumption over the Dwyer's instream flows. As with many other basins, such as the Ganges/Brahmaputra and the Gambia, the measurement of the river's discharge to the ocean is not easy. It is carried out at the first feasible point upstream from the coast.

Of course, the one-third rule requires an effective system of abstraction licensing and drought management. Instream flows are valued by the people and their government not merely because of their economic output, already referred to above, but also because of the intrinsic value of the riverine and coastal ecology.

Table 8.2 Population, productivity and output change over time.

	Rate of growth of population in per cent per year					
Rate of growth of output per head in per cent per year	0	1	2	3	4	5
0	1.000	1.010	1.020	1.030	1.040	1.050
1	1.010	1.020	1.030	1.040	1.050	1.061
2	1.020	1.030	1.040	1.051	1.061	1.071
3	1.030	1.040	1.051	1.061	1.071	1.082
4	1.040	1.050	1.061	1.071	1.082	1.092
5	1.050	1.061	1.071	1.082	1.092	1.103

Note: The statistics above, lying between 1.000 and 1.103, can be used as multipliers to show how output changes between any two successive years. The decimals give the rate of change of output. For example, 1.051 implies a rise of 5.1 per cent in output between any two years. At this rate of growth, output would double in 14 years.

It has already been seen that the catchment's population is increasing. Output per capita is also rising. This combination of growth both in population and economic productivity, associated with the rise of world capitalism, is the source of deep unease amongst the professional staff of the Dwyer Catchment Authority. The Authority has used Table 8.2, and data from the Ministry of the Economy and Population, to estimate the catchment's economic output in 2025. After assuming that the ratio of out-stream water use to basin output is constant, the Authority also has a derived estimate of total water use in 2025 (row 33 of Table 8.1 for the scenario year 2025). The Authority anticipates that, for the first time in its history, the catchment will move into a water deficit; one or more of the four necessary conditions for surplus will have been breached.

8.3 REDEMPTIVE OPTIONS (I)

The Dwyer Catchment Authority responds by considering and starting to implement redemptive options that moderate the situation *but that do not prevent entry into water deficit*. There are six such adaptive measures.

The first option is to amend the 1847 Act so as to lower the measure of environmental needs below the 33% floor. The decline in environmental flow (and the ecological and economic effects that ensue) is accepted by the catchment authority (if not by the environmental movement) as 'a price worth paying' to maintain the growth of population, economic productivity and total output.

The second option is to promote the reuse of abstracted water as a means to cut the rate of abstraction. Reuse can take place *within* a single institution (such as a sugarmill) or it can occur *between* institutions as when urban waste water is reused for irrigation purposes. Reuse may have real advantages, such as lowering abstraction costs or by reducing waste water treatment costs. But there is no reason to believe that reuse reduces the consumption of water supplied to and used by households, firms, etc. One can take a small sugar mill as an example. In the absence of reuse it abstracts 1000 m^3 per week, 150 m^3 of which is consumed in evaporative losses leaving 850 m^3 returned to the river. In the case of reuse only 500 m^3 are abstracted on Monday and by Wednesday 75 m^3 are consumed and replaced from storage. By Saturday the reuse stage is complete and a further 75 m^3 consumed, once again to be replaced from storage. The reuse option has abstracted less than in the absence of reuse (650 m^3 including storage recharge), but there is no difference in total consumption. Since reuse has not brought with it a fall in consumption it has not served to protect the instream flow.

The third option is rainwater harvesting. These projects can be successful at the local level; they are specific forms of water capture closely related to surface water abstraction. As with reuse, there is no fall in consumption and no protection of the instream flow.

The fourth option is to import water from another catchment. This permits the sustainable management of the Basin's aquifers, sufficient outstream supply to match the economic demand for water, satisfaction of the economic demand for food and continued protection of the Dwyer's environmental flows.

The fifth option is the desalination of sea water and other saline waters. The power requirements are considerable but recent falls in the factory-gate price per cubic metre make this option an interesting one for the Dwyer Catchment Authority. The use of sea water itself, requiring dual supply infrastructures, is also of interest to the Authority in some sectors. Hong Kong sets an example in this respect.

The last of the six options that the Authority may take to win hearts and minds as the entry to water deficit begins is to import food. This permits administrative or legal restrictions on the scale of irrigated agriculture such that there is now sufficient water for all other economic demands. A beneficial allocation multiplier exists here. If water use for irrigation and for urban purposes divides in the ratio 70:30 (Shiklomanov 2000: Table 5), then a 15% transfer of agriculture's total creates a 35% increase in the urban total. The consumption multiplier is even greater. If, water consumption by the irrigation sector and the urban sectors divides in the ratio 93:7 (*ibid.*), then a mere 5% transfer of agriculture's total permits a 66% increase in the urban total.

Each of these six options above needs to be reviewed. The first redemptive option reduces environmental flows below what had been considered the minimum acceptable, breaching condition 4 for water surplus (see Section 8.1). The second option is welcomed on economic grounds but makes no contribution whatsoever to defending the basin's water surplus. Nor does the third option. The fourth measure places reliance on the precipitation of other catchments rather than precipitation in the Dwyer River Basin. Parallels spring to mind with Israeli pumping of Lake Tiberias, the plans to divert water from the Ebro River to south-east Spain, and the south–north transfer in China (Pearce 2003). The fifth measure is noteworthy but, like option four, only compensates for a shortfall in the Dwyer's precipitation. The sixth option is in breach of condition 2 for water surplus. Moreover, if the basin's exports are insufficient to finance food imports, condition 3 is breached. The Dwyer places reliance on other catchments' food production and, indirectly, their precipitation.

It is worth noting that the argument above relies (implicitly) on a distinction between the terms 'reuse' and 'recycling'. Reused water is waste water and irrigation drainage that, *prior to its return to the hydrological resource*, is captured and used again (perhaps repeatedly) for the uses listed in Table 8.1.

Recycling refers to water that is abstracted, on the supply side of the hydrosocial balance, used by households, industry etc., and then the fraction that is not consumed flows directly back to the catchment's rivers, aquifers and lakes. The importance of recycling to hydroeconomists is that it augments the hydrological resource from the point at which the recycling occurs. The negative characteristic is that recycled flows may pollute the resource. The proportion of water used that is consumed, and therefore is unavailable for recycling, varies between categories of use. Therefore the relative shift of use between Table 8.1's categories of use will change water consumption for a given volume of use. The example above of the beneficial multiplier illustrates this.

8.4 REDEMPTIVE OPTIONS (II)

The Catchment Authority continues by considering redemptive options that *do* prevent entry into water deficit. Once again, there are six such adaptive measures. They are numbered 7–12 to maintain continuity with redemptive options (I).

The seventh option is to reduce the rate of growth of population. Other things being equal, this will reduce the rate of growth of output and thereby the rate of growth in outstream water use. A fall in demographic increase within the Dwyer River Basin would be achievable if its women become more powerful in the determination of the number of children conceived, if access to contraception is facilitated, if social provision for elderly persons is strengthened, and if limits are placed on net inward migration. China's population policy since 1949 has contributed enormously to the impact its society has on the available water resource, in comparison with the population totals it would now have in the absence of that policy.

The eighth option is to reduce the rate of growth in output per person and thereby the rate of growth in output and outstream water use. In the Dwyer River Basin this would be difficult to defend because there is extensive poverty in the catchment. In basins with higher per capita incomes it would require a fundamental change in cultural values and economic organization, a farewell to capitalism's relentless search for output growth founded on market expansion.

In the case of options seven and eight one sees that these require action that clearly falls outside the field of water resources management. But this is not true of our final four options. The ninth adaptive measure is to increase outstream water's productivity in terms of value-added (in US$) per cubic metre of water consumed. It is consumption, not use, which reduces the volumetric flow of the Dwyer. Water productivity augmentation is a subject of discussion, research and implementation throughout the catchment with respect to all the outstream users. The greater is water's productivity, the lower the volume of water consumed for any given level of future output. It has already been seen above that at the global scale 93% of all consumption of water withdrawals takes place within the irrigated sector. In the Dwyer Catchment it is irrigation yields (in terms of output per hectare, output per cubic metre of water used and output per cubic metre of water consumed) that are considered as the key variables in respect of option nine. The positive link between food prices and crop yields is of especial interest (Berkoff 2003).

The tenth option relates to the economic demand for water use under condition 2 for water surplus. The economic demand for water is analysed by economists as a functional relationship between the price of water (money charged per unit volume) and the total volume users wish to purchase at that price. A higher price is usually associated with a lower volume purchased. If it is possible to extend water pricing in the Dwyer River Basin as well as to raise the current level of prices, then the volume of water purchased by households, farmers, companies and other institutions will diminish.

Differential pricing for users that recycle a high proportion of their supply helps reduce water consumption. The introduction of pricing, particularly in the irrigation sector, can be immensely difficult (Merrett 2002b). A subsidiary arrangement under option ten is to facilitate an urban market in farmers' water-rights, taking advantage of the allocation multiplier.

A positive spin-off from pricing is the stimulus it gives to higher water productivity. A negative spin-off would be a contraction of irrigation output leading to food imports that the Dwyer population is unable to finance out of non-farm exports. Water pricing can also be deployed to moderate abstraction and use on a seasonal basis; this can be of value when precipitation, evaporation and river flows vary strongly during the course of the year, as is the case with the Dwyer catchment.

The eleventh option is to reduce the evaporation from dams and reservoirs. A deep dam has an advantage here over several shallow dams with the same total storage capacity. Artificial aquifer recharge, substituting for surface storage, is of great relevance in those areas of the Dwyer River Basin where the rate of evaporation is high because of heat, wind and low humidity.

The twelfth and last option, for society as a whole, is to establish water resource institutions (public, private and cooperative) that have an effective leadership, are properly resourced, of considerable expertise, strongly committed to the agenda of a sustainable society and, in the case of the Dwyer Catchment Authority, with the necessary regulatory powers (Abernethy 2003; Merrett 1997: 143–62).

8.5 FROM SURPLUS TO DEFICIT

At this point it is necessary to consider whether or not it is important that one catchment is in deficit and whether or not it is of any significance that a second catchment is *moving* from water surplus towards water deficit. The issue can be approached by re-stating the implications for a river basin of finding itself in water deficit. These are one or all of the following:

1. The basin is pumping its groundwater at an unsustainable rate.
2. There is insufficient outstream water to meet economic demand.
3. The population has to import water, or food that it is unable to pay for from its exports of goods and services.
4. The basin's citizens must accept the economic and environmental losses following from its river diminishing in volume.

The nightmare scenario is a basin in which groundwater is being exhausted, households, farmers and other actors cannot purchase the water they require, food imports cannot be paid for, making the basin dependent on powerful allies, and the river has been destroyed.

Consider too the global catchment – the set of all the world's river basins. As the world's population increases from six billions in 1999 to some eight billions in about 2025, and as world economic growth maintains an even higher rate of expansion, is it not the case that the global catchment itself, whilst still in surplus, will embrace ever more individual river basins that are in deficit? As the global catchment's surplus diminishes, will not the situation of specific catchments in deficit become ever more precarious? Moreover, since water flows uphill to wealth and power, will not the social classes that have neither wealth nor power feel these deficits most intensely? It is this prospect that is the basis of the importance of water deficits and the significance of the shift of river basins from water surplus to deficit (Lundqvist 1998).

8.6 FRAMEWORK, THEORY AND EMPIRICAL STUDIES

It may be that this approach to catchment deficits and surpluses will be of interest to the international community of researchers working on water resources and their management. If that is so, the analytic framework requires testable hypotheses. The *general hypothesis* is: 'Catchment X is in water deficit', where 'Catchment X' is the specific river basin to be researched.

To test the general hypothesis, the most basic information on the catchment that is required for a catchment in a specific year is:

- Total precipitation.
- Effective precipitation.
- Net precipitation.
- The sustainable rate of abstraction of groundwater.
- The actual rate of abstraction of groundwater.
- The performance of water suppliers in meeting the economic demand for water from households, agriculture, mining, manufacturing, construction and the services sectors. The hydrosocial balance of the catchment will prove useful here.
- The ability of consumers to exercise their economic demand for food. Where such food is imported, are these imports purchased at the international market price and is this financed from the catchment's foreign exchange earnings?
- The quantitative flow of the river at the point of measurement of its discharge to the ocean.
- A statement of the defined minima for the river's instream flows and the actual flows that correspond to these environmental standards. If no environmental standards are in place, the Dwyer's one-third rule may be used as a substitute.

In the light of this information, if the four conditions for water surplus of section one are met, and if this is possible without reliance on desalination or water imports, then the general hypothesis is falsified and the catchment is in surplus. If not all four conditions are met, but only as a consequence of water exports, the catchment again is in surplus, although it may suffer the symptoms of deficit. The Jordan River Basin may be an example.

Where the general hypothesis is *verified*, an appraisal of redemptive options 7–12 is overdue. Where a catchment is in deficit, one can hypothesize that, in comparison with basins in surplus, net precipitation per head is low, or output per head is high or water productivity is low – the *causal hypothesis*.

Finally, if a number of Ph.D. students, professional researchers and institutions adopt this approach, test the hypotheses and appraise the redemptive options, a splendid panorama opens for interchange in respect of framework, theory, and empirical studies.

8.7 CONCLUSION

The paper introduces the concepts of catchment water surplus and catchment water deficit. It describes how a shift from surplus to deficit can take place as population and output per head in the catchment increase. Twelve redemptive water policy options to manage or halt this shift are set out and the significance of such shifts at the global scale are considered. The paper ends by formulating two testable hypotheses derived from the general theory and invites professional colleagues to pursue, within this paradigm, investigations of specific river basins.

The material in this chapter originally appeared in: Catchment water deficits in the 21st Century. *Water Policy*, forthcoming 2004.

9

The urban market for farmers' water-rights

9.1 INTRODUCTION

In the field of water resources management a widely held belief exists that allocation stress is to be found in many parts of the world and is set to become more intense in the future because of global population growth and climate change (Meinzen-Dick & Rosegrant 2001a). Allocation stress refers to access conflicts between the agricultural, domestic, industrial, urban service and environmental uses of outstream and instream water flows. If the belief that allocation stress will intensify is well founded then, because of the dominant role of irrigation water use at the global level, it is imperative to explore the possibilities of reducing farmers' use of water or, at the very least, of slowing its growth (Merrett 2002a).

One process by which the scale of irrigation is reduced occurs when farmers choose to sell their abstraction rights (or other rights to access water) in perpetuity to actors that apply these released flows in towns and cities for household, manufacturing and urban service uses. It is this market that is the subject of the present paper and the term 'market' is used here to refer to those institutions that provide the context for the purchase and sale of commodities. A beneficial allocation multiplier exists here; if water supply for irrigation and for urban purposes divides in the ratio 70:30, for example, then a 15% transfer of agriculture's total creates a 35% increase in the urban total. Note that the paper does not address environmental and hydropower needs or the transfer of abstraction rights between farmers. Neither does the paper review the broad-ranging arguments for and against re-allocating water from farming to the urban sphere (*vide* Merrett 2002a: 148–82, Rosegrant & Ringler 1998).

The objectives of the paper can now be stated. These are: first, to set out an orthodox microeconomic account of a water-rights market in equilibrium; second, to ascertain whether that approach can deal with a number of empirical complexities in real-life trading.

The final introductory comment concerns the conceptual distinction between water and water-rights. Water is a collection of molecules, each of which consists of two atoms of hydrogen bonded to one atom of oxygen. A water-right is a legal claim to abstract or otherwise access water. You can float on water but not on a water-right. You can abrogate a water-right but not water. This paper addresses water-rights markets, not water markets.

9.2 THE URBAN ACTORS' DEMAND FUNCTION

We begin with a defined region composed of a metropolitan area located within an agricultural landscape in which farmers abstract surface water and pump groundwater for irrigation purposes. Abstraction takes place within a framework of rights that are formalized in national laws, or are recognized by customary practice, or exist in situations of legal pluralism (Bruns & Meinzen-Dick 2001: 1).

The city already accesses water from a variety of sources, as the hydrosocial balance shows. These sources include:

1. Rainwater harvesting in cisterns by households.
2. Groundwater and surface water abstraction by the metropolitan water utility, urban developers and by manufacturing industry.
3. The utility's import of water from a neighbouring region.
4. The manufacturing sector's reuse of its own waste water.

The population, gross domestic product, income and size of the city are all growing – as is the demand for water. Consequently the metropolitan water utility as well as the enterprises that capture their own water needs (together the *urban actors*) are seeking to expand their separate water supply capacities. In particular, the water utility knows that supply expansion requires capital finance and infrastructural construction. This will push up the firm's overhead costs and prime costs and thereby total costs (Merrett 1997: 32–5). But the utility is sure that if the average total cost per cubic metre of water delivered remains close to its existing level, the price charged to customers will be one that they are willing and able to pay.

The utility and those companies self-supplying their own water requirements begin to take an interest in the abstraction rights enjoyed by farmers in the surrounding countryside. They know that these rights can legally be transferred to urban actors. For any single project the urban actor would have to pay a lump sum to the farmer for the transfer of the abstraction right, as well as meeting accompanying transaction costs. In addition, the urban actors would have to make capital investments in headwork and network infrastructures in order to capture the water they are now permitted to abstract and to deliver it to the city. The quality of water from any specific abstraction point would likely impact on headwork treatment costs, particularly with respect to the water's potability.

So what is the maximum payment an urban actor would be willing to pay to the farmer for the transfer of his abstraction right, that is to say, what is the urban actor's

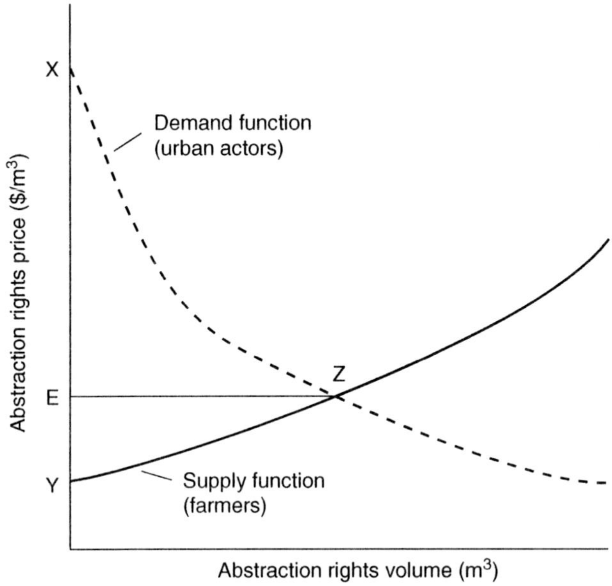

Figure 9.1 Demand and supply functions for farmers' water-rights.

maximum bid price? One answer is that it would be equal to a cash sum that when added to transaction costs and to the accompanying capital expenditure on infrastructure would generate an average total cost per cubic metre of the same order as the existing level of average total cost. A second answer, a stronger response, is that the maximum bid price plus the other capital spending would not produce an average total cost per cubic metre higher than the average total cost of the best alternative supply scheme (Kemper & Olson 2000: 352). For example, such alternatives might be an increased capacity of the utility to import supplies from another region or the expansion of a manufacturing company's waste water reuse facility. In the calculation of average total cost, the volume used as the divisor should be that delivered to the customer. In this way storage and distribution losses are properly accounted for. This point is made strongly by Schiffler (1997: 368).

There may also exist urban institutions requiring abstraction rights for whom no feasible alternative supply source exists. In this case the maximum bid price would be that which, after the capital payments associated with the rights transfer, still leaves the investment project with an acceptable financial rate of return. As an alternative to supply expansion, the utility could also consider a tariff rise.

In any given year there are a number of urban actors each with one or more potential abstraction rights purchases in preparation but in which the final purchase price has not yet been legally agreed. For urban actors as a group, a function of the type represented in Figure 9.1 can be conceived to exist. The vertical axis measures, for urban actors, their maximum bid price per cubic metre in 2003. The horizontal axis measures the abstraction rights volume purchased. We have a standard demand curve. For a modest addition to the urban actors' abstraction capacity, high prices are bid. This is because such additions are necessary to meet clearly identified needs over, for example, the years 2004–2008 (Rosegrant & Binswanger 1994: 1617). Much lower prices are bid for a large volume transfer of rights because they are not required in the short-medium term.

9.3 THE FARMERS' SUPPLY FUNCTION

The farmer who sells his abstraction right is here assumed to relinquish his entire legal claim on water use for irrigation purposes. Hereafter 'farmer' can be taken to refer to a farming family, an agribusiness, and an irrigation district or farmers' association that possesses water-rights. As Hearne and Easter point out (1995: 41): 'These rights can be stipulated by volume or by percentage of river ... flow. But in areas where water supplies are highly variable, it is necessary to designate how water is allocated during times of scarcity'. Rosegrant & Gazmuri Schleyer (1996: 270–1) make the same point. The farmer who sells his abstraction rights reverts to cultivation activities that are feasible on a rainfed basis. This may make it necessary that he incurs some additional capital expenditure and it will certainly entail a loss of income from produce sales, because of the fall in yields under the new conditions in which crop water requirements cannot be met by irrigation.

This reduction in farm income (net of any reduction in prime costs) plus any new capital expenditure required (including transaction costs) must be compensated through the price that the farmer receives from his sale of the abstraction right. He sets his minimum release price equal to new capital expenditure plus a multiple (here referred to as m) of the annual loss of net farm income. (There is a parallel here with the payback period required by some investors in order to go ahead with a new investment project.) From the point of view of theory, the fact that the farmer may set m equal to (say) 10 years or less demonstrates his discounting of future returns. The value of m can be expected to vary with crop type; and estimated annual income loss may include a risk premium for cultivation dependent entirely on rainfall.

Figure 9.1 represents the supply function. The abstraction rights volume offered by farmers at the lowest price is small but increases at higher prices. There is an equilibrium price E at which the market is cleared. At this price the abstraction rights volume sought by the metropolitan water utility and other urban enterprises is precisely equal to the volume farmers are willing to release. Note that at the price E urban actors' total expenditure on the acquisition of abstraction rights is lower by the area XEZ than the demand price function (built up from individual urban actors' maximum bid prices) shows they would have been prepared to pay. At the same time farmers' total income from the sale of their abstraction rights is greater by the area YEZ than the supply function (built up from individual farmers' minimum release prices) shows they would have been prepared to accept. Everybody's happy. Economists will recognize that the supply and demand analysis used above is of a type that most basic orthodox economics texts deploy.

9.4 THE LIMITS TO THEORY

The previous sections have presented a simple theory of a rural–urban market in abstraction rights. But simple and elegant societies to which such a theory might apply do not exist. The point has come in the argument where it is necessary to muddy these crystal waters with real-life impurities if our market theory is to have application to actual transactions.

9.4.1 Time-scale

The analytic exposition is based on the trade of water-rights 'in perpetuity'. However, Rosegrant & Meinzen-Dick point out:

> While tradable water rights should be permanent, or very long term, to ensure security of the right, transfer of water rights need not be permanent: water rights can be leased for a season, a year, or many years. (Rosegrant & Meinzen-Dick 1996: 47)

The present paper does limit itself to long-term or permanent trading. In the case of the transfer of water-rights, the shorter the lease, the closer one approximates a market in water itself. Rosegrant & Gazmuri Schleyer (1996: 276–7) give examples from Mexico both of a one crop season transfer in Monterrey and a 50-year trade in Guanajuato. The Mexican Comisión Nacional del Agua permits concessions and grants for periods from 5 to 50 years, with terms exceeding 30 years the norm, 'to ensure security of the water right' (*ibid*. 269).

It is not wise to present the theoretical approach above as a basis for all trades and transfers, whatever their length. From the point of view of both the farmer and the urban actor a one-year lease, for example, is an asset utterly different from a permanent sale. Only in the latter case does the farmer give up for all time the use of this flow of water for irrigation; and only in this case does the urban actor secure for the indefinite future its water source.

However, it must be admitted that the short term may be the prelude to the long term. Haddad (2000: 95–116) has described a fallowing agreement in respect of about 118 million cubic metres of water per year for the 2 years ending in July 1994 between the Palo Verde Irrigation District and the Metropolitan Water District of Southern California. In 2002 the two parties were negotiating a contract to fallow up to 29% of the irrigated farmland for a 35-year period (A. Quist, pers. comm.).

9.4.2 Concessions

The general theory is framed in terms of a market in abstraction rights. But the use of water by farmers is also made possible when they enjoy a legal right to access water flows from an organization that itself is the abstractor. The obvious case is where a public irrigation authority assigns farmers the right to draw off water from a canal. The model of Sections 9.2 and 9.3 can incorporate such 'concessions' (as they are sometimes called) without amendment and permits me to refer to it as a theory of the market in tradeable water-rights rather than one limited to trades in abstraction-rights. Trading in the surface water concessions of a public authority may also often fall within a leasing model (see Section 9.4.1 above).

9.4.3 Part-sales

Here we consider the situation where the farmer sells only a fraction of his water-rights, retaining the rest for irrigation on a smaller scale. The theory incorporates this with no difficulty. The partial sale of the farmer's water-rights is still represented by the supply function; the specified volume is smaller but the release price per cubic metre may be little changed. A cultivator may choose to use the payment he receives to invest in water-saving technology for the areas he continues to irrigate (Gazmuri Schleyer & Rosegrant 1996: 41). However, from the point of view of the urban actor, to attain a targeted volume from a group of partial sales rather than a single, high-volume trade implies increased storage, distribution and transaction costs.

9.4.4 Sale of land

Here we have a situation where the farmer sells not only his water-rights but also his farm and the land on which it is set. With respect to the cultivator's release price, he now has to face a future in which he derives no income stream from his property. The minimum release price is now set at his transaction costs plus a multiple n of the total annual surplus, that is, the farm's turnover less the sum of prime and overhead costs. The minimum release price for the joint sale of land and water-rights exceeds that where the sale is only of the water-rights, because in the former case alone is all farming income lost. A farmer receiving a capital sum for the sale of his land and water-rights may well use the money to clear any debts he may have.

With respect to the maximum bid price of the urban actor, he now purchases two assets – the farm property as well as the water-rights. There may be significant advantages here. The purchaser may believe that the risk of legal challenge to his water-rights is considerably diminished by virtue of his land ownership status. Moreover, access to his new headwork and network infrastructures is now entirely within his control. Finally, parts of the farm unrelated to the abstraction and distribution process can be sold off, immediately or in the future. As a result the urban actor's bid price for the twin assets enjoys a premium over the bid price for the water-rights alone.

These remarks on urban actors' position can be put more forcefully. In many countries water-rights and rights to land are indissolubly linked; in order to access the former it is necessary to buy the latter. In such cases a metropolitan area pursues its search for water by means of land purchases within its region. The most famous example is that of Los Angeles and Owens Valley (Reisner 1993: 59–96).

These remarks on concessions, part-sales and the sale of land address some of the complexities of the real world. They also significantly widen the scope of the theory yet leave the theoretical approach more or less intact.

9.4.5 The legal context

For the theory to have an application to inter-sectoral markets, a strong condition is the existence of secure water-rights within the ruling legal framework. These must be rights for the farmer to access water (by abstraction or concession) from an aquifer, river, lake or canal at defined times and in defined quantities and they must define how access will be reduced in times of drought. Rights may take the form of a share in stream flow. The legal system must also recognize that the farmer is permitted to transfer this right to an urban actor.

In the absence of such a legal basis a market may still exist, but it will be arcane. The current market in Mexico's Baja California Sur is an example (E. Troyo: pers. comm.). In such circumstances urban actors' bid prices will be lower because of the perceived risk that they may in future lose the access to their water source following a private law suit or through a government ban. In fact, Bruns & Meinzen-Dick (2001: 3) suggest that '... industries rely on their economic and political power to obtain water from farmers, legitimately and illegitimately. Institutional frameworks for carrying out reallocation through voluntary agreements among users are usually non-existent'. They suggest a dual approach to providing a legal basis for trade. Where transactions are likely to be sparse, the framework would simply enable transfers to be negotiated and allow challenges by those who feel they may be harmed. In other cases laws and agencies would establish a cadastre, introduce requirements for public notice, provide expert assessment of third-party impacts (see below) and assist the less educated and powerful in the trading process (Bruns & Meinzen-Dick 2000). In this context it is suggested that:

> ... the establishment of markets in tradable property rights does not imply free markets in water. Rather, the system would be one of managed trade, with institutions in place to protect against third-party effects and possible negative environmental effects ... (Meinzen-Dick & Rosegrant 1997: 317)

9.4.6 Third party effects

When a farmer irrigates, some proportion of the water used is recycled downstream or back to the aquifer. Shiklomanov (2000, Table 5) estimates at the global scale that 70% of agricultural withdrawals are consumed and the remaining 30% are recycled. Following the sale of an abstraction right or a concession to an urban actor, the water is physically transferred to the water utility, to a residential developer or to an industrial enterprise. As a result the recycling (at an increased but perhaps polluted volume) takes place downstream of the new users – households, manufacturing or urban service companies. The upshot is that the third party interests of farmers and other actors downstream of the original point of irrigation are affected. They lose the recycled flow they had enjoyed prior to the inter-sectoral transfer. But note that some farmers may experience greater recycled flows if the city in question is upstream of the farm selling its water-rights.

A third party effect also exists when an urban actor purchases a water right not utilized fully by the farmer and then exploits that right 100%. The downstream flow is correspondingly reduced. As Kemper (2001) writes:

> ... [the downstream neighbours] may go to court to prevent the sale, thereby causing high transaction costs, or courts may not be available and they will have to accept the loss. A mechanism is therefore needed to either negate third-party interests or to mitigate the impacts of water trading on the different stakeholders, including an effective conflict resolution system.

So the approach to the return flows issue can stand at two extremes. In one case, there are no private rights to return flows; government retains them. Here the sale of a water-right by a farmer to an urban actor that results in a diminution of return flows cannot be challenged by the downstream party. This has long been the case, for example in Chile, in Mexico and in the Northern Colorado Water Conservancy District of the USA. At the other extreme farmers and other instream water beneficiaries have a right to the return flows they receive and therefore can seek to block a development that reduces these flows or, at least, to be compensated for their loss. The transaction costs that must be met in agreeing a water-rights transfer are consequently higher (Rosegrant & Binswanger 1994: 1619, Gazmuri Schleyer & Rosegrant 1996: 37).

There is another source of third-party effects. In section one it is stated that this paper does not review the broad-ranging arguments for and against re-allocating water from farming to the urban sphere. However, it is clear that a large-volume reduction of irrigation flows can have powerful, negative impacts on a rural economy in which farming and agriculture-based services, manufacturing and transport are major employers. Water-rights trades can 'dry up' the economy itself. In New Mexico, for example, some *acequías* have gone to court to block sales because they threaten the community and local livelihoods (Meinzen-Dick, pers. comm.).

9.4.7 The market

The next real life complexity reviewed here concerns the market within which inter-sectoral trades take place. A physical market-place (such as a particular café) may exist, or transactions simultaneously involving a number of farmers and urban actors may occur through an open, printed or digital form. Here potential buyers and sellers can meet and gather information on the rights that are in trade.

In the absence of such a market, deal-making is likely to be on a bilateral basis. A specific urban actor seeks out a farmer who may be willing to sell his water-rights and they conclude an agreement in the partial or complete absence of information about other potential purchasers and vendors. Here there is a strong likelihood of asymmetric information with the urban actor better informed about the farmer's negotiation position than vice versa. The haggling process tips in favour of the water utility or the urban enterprise. Karin Kemper writes (2001):

> Socio-economic asymmetries need to be considered because of the different types of water users, who will vary in their educational background, culture, and economic power. When water rights were allocated and made tradable in Chile, electricity companies bought up a large number of them to be held for future use, to the detriment of smaller users who at the outset did not understand the implications of their selling the rights.

In regions where bilateral deals are dominant *there is neither a market nor an equilibrium price.* Hearne & Easter's valuable study on Chilean water rights markets clearly indicate that the number of intersectoral transfers was low. This was true of the upper Maipo valley, the Azapa valley and the Elqui valley. In the Limarí valley 'transactions are fairly frequent', but the text suggests these are predominantly transfers within agriculture (1995: 32–6, 50–3). In contrast a study of the water register of Santiago County in Chile for 1993–4 indicated there were some 500 permanent, agriculture-urban trades. They equalled only about 0.1% of the total water rights held in the area (Gazmuri Schleyer & Rosegrant 1996: 42–3). In fact the best quantitative expression for the importance of water-rights trading, past and present, to the water requirements of a city is the volume of water comprised in such trades over time divided by the total volume of water the city receives in the current year.

9.4.8 Transaction costs

These have already been referred to several times and are fully incorporated into the general theory. The legal context strongly influences the transaction cost associated with any rights transfer. The greater the cost, the less likely it is that a deal will be struck. Transaction costs include the time and money expended on searching for a buyer or seller of water-rights, negotiating and legalizing a contract, validating the legal ownership of the water-rights, payment of any government tax on transfers, and enforcing the contract. In his *Rivers of Gold* Haddad (2000) makes clear in a set of fascinating and detailed case studies from California that the management of risk lies at the heart of transaction costs. Note that these costs are those incurred in the transfer of water-rights, not those incurred in the physical capture and distribution of water. The latter is a commonplace supply cost (Merrett 1997: 5–40).

9.5 CONCLUSIONS

Allocation stress and the policy responses to the dilemmas it poses will be a prime source of debate in the field of water resources management during the next decades. In the broad sweep of world history, the transfer of water-rights from farmers to urban actors has been one form in which allocation stress has been managed. This has taken place through the sale of such rights or their seizure by *force majeure*; in both cases such sales have frequently been accompanied by the sale or seizure of riparian land.

This paper set out to accomplish two objectives. The first was to set out an orthodox microeconomic account of a water-rights market in equilibrium. This was done by representing urban actors' requirements for farmers' water-rights as a demand function based on each urban actor's individual maximum bid price. The rural response was represented as a supply function based on each farmer's individual minimum release price. The second objective was to ascertain whether this neo-classical approach could deal with a number of empirical complexities in real-life trading.

The conclusions are that the theory capably handles not only abstraction rights but also concessions, part-sales, accompanying sales of land, and transaction costs. However, the theory's applicability requires a strong enabling condition: that the legal system fully incorporates such transfers (including a clear stance on third party rights) and that an open, well-informed market exists.

But the empirical material cited shows that the absolutely predominant form of transaction is the bilateral deal. Bilateral transactions predominate because the total number of annual sales in a defined market area is usually small, eliminating the possibility of a thriving market process involving multiple sellers and multiple buyers. Rather than a competitive and open market we typically see only one or a small numbers of sellers, that is, a rural monopoly or oligopoly. Similarly we find only one or a small number of purchasers, a monopsony or oligopsony. The conclusion is that a theory of the neoclassical type does not well represent actual social processes. To understand these processes, research is required into one-off, scattered, fragmentary, arcane deals, often of dubious legality and marked by information asymmetries.

However, it seems that the concepts of a farmer's minimum release price and an urban actor's maximum bid price are still appropriate, as is the account of how these prices are shaped. Similarly, a rural monopoly–urban monopsony approach can also deal not only with abstraction rights but also concessions, as well as part-sales, land sales and the transaction costs structured by formal rights, customary practice or legal pluralism.

The material in this chapter originally appeared in: The urban market for farmers' water rights. *Irrigation and Drainage* 52(4): 319–326, 2003.

10

The demand for water: four interpretations

In Geometry (which is the only science that it hath pleased God hitherto to bestow on mankind) men begin at settling the significations of their words; which … they call Definitions.

Thomas Hobbes, *Leviathan*, 1651

10.1 INTRODUCTION

An extraordinary characteristic of the current management of global water resources is the breadth of disciplines on which it draws. To name but a few one can cite agronomy, civil and hydraulic engineering, economics, environmental science, geology, human and physical geography, hydrology, land-use planning, meteorology, political science and sociology. Most readers of this paper will have received a professional training in at least one of these subjects.

Each of these disciplines has its own language and not even a Darwin, a Keynes or a Newton could dream of mastering all of them. Yet every historian of science recognizes how critical is the language of a discipline to its success and how vital the development of the language of science is to science's progress (Kuhn 1970). As Dow writes (2002: 5) 'It is normal in scientific thought for terms to change meaning over time'.

A feature of water resource debates at the present time is the wide application of the term 'demand'. It may be that civil engineers first gave it real weight in water resource planning. Civil engineers saw themselves above all as responsible for water supply and therefore (this is an hypothesis) assigned the term 'demand' to refer to those outstream flows that began at the point where engineers put into supply the water they had abstracted from lake, river and aquifer.

What is not in doubt is that 'demand' is a concept applied by professionals and policy-makers from many disciplines. It is the objective of this technical note first to show that the word has a number of quite distinct meanings that it is unhelpful to confuse, and second to show that distinguishing these meanings from one another will strengthen planning practice.

10.2 THE USE OF WATER

The thesis presented here is that there are four demand-side concepts with respect to outstream water: use, consumption, need, and economic demand. *Water use* is the

Table 10.1 The hydrosocial balance for a defined region in 2003.

Row	Title	Base year volumes (Mcm)
4	**Categories of Supply**	
5	Rainwater collection	a
6	Groundwater abstraction	b
7	Surface water abstraction	c
8	Desalination	d
9	Imports of water from other areas	e
10	Internal reuse of wastewater	f
11	External reuse of wastewater	g
12	Net fall in water abstracted and stored	h
13	**Total Gross Supply**	**j**
14		
15	**Categories of Supply (Negative)**	
16	Supply leakage and evaporation	k
17	Exports of water to other regions	l
18	Net rise in water abstracted and stored[1]	m
19	**Total Negative Values**	**n**
20		
21	**Total Net Supply**	**j − n**
22		
23	**Categories of Use**	
24	Households	o
25	Agriculture	p
26	Mining	q
27	Manufacturing	r
28	Public services	s
29	Private services	t
30	Other uses	u
31	**Total Gross Use**	**v**
32		
33	**Categories of Use (Negative)**	
34	Leakage and evaporation during use	w
35	Net rise in water received and stored[1]	x
36	**Total Negative Values**	**y**
37		
38	**Total Net Use**	**v − y**

Note: [1] A net fall (rise) in storage of water has a positive (negative) value.
Source: Adapted from Merrett (2002a) Tables 7.1 and 7.2.

first and easiest concept to grasp. A generic way of tabulating the regional users of outstream water is in terms of households, agriculture, mining, manufacturing, public services, private services, and other uses. In each case we are referring to the quantitative flow per unit time of water arriving at a user's property (a dwelling, a farm, a mine, a factory, a hospital, an office block, etc.). That flow may be used beneficially or it may be wasted in various ways. Directly or indirectly, water use is the basis of human life and it makes up one of the two building blocks of the hydrosocial balance in a base year (such as 2003) or in a future scenario year. The generic form of the balance is set out in Table 10.1.

10.3 THE CONSUMPTION OF WATER

During the course of using water, some part of it returns to the hydrological cycle through evapotranspiration; this is the second concept of interest to us – *water consumed*. What is left after the use process is complete is waste water and irrigation returns. These flows are reused internally or externally by user groups, or recycle as run-off to surface water, or recharge groundwater.

The consumption of water is of particular interest because the ratio of consumption to waste water plus returns varies significantly between types of use (Shiklomanov 2000). To take an example, a Colombian sugar-mill in the Cauca Valley that internally reuses its own waste water to avoid abstraction charges seeks a low level of consumption. In contrast, an English farm in the Anglian region irrigating containerized trees has 100% consumption – all its water is evaporated or transpired. The ratio of consumption to returns has generated a major debate on irrigation efficiency (Perry 1996).

10.4 THE NEED FOR WATER

The third of our concepts is *water need*. For households water need is a social, cultural and health-related concept; it refers to desirable or recommended levels of use for drinking, cooking, personal washing, disposal of urine and faeces, cleaning and washing in the house and watering of a kitchen garden. In the farm sector water need refers to net irrigation requirements. For mining, manufacturing and the services sectors, their need for water is technically and institutionally determined. So the central difference between water use and water need is that the former represents an actual flow to users whereas the latter represents a desired or recommended flow.

Returning to the hydrosocial balance, estimated water need in a baseline year can be compared with water use, possibly leading to recognition of unmet need. For a scenario year, water need can provide the starting-point for planning an expansion in supply.

10.5 THE ECONOMIC DEMAND FOR WATER

The fourth concept is the *economic demand for water*. For economists the concept of 'demand' refers to the relationship, at a given time and within a defined market, between price per unit of any product or service and the quantity in each time period that users are willing to purchase at each price. The demand function is conventionally represented graphically with price on the vertical axis and quantity on the horizontal axis, showing the difference in quantity purchased at each price (e.g. Merrett 1997: Fig. 4.1).

The cultural and economic context of any demand function is referred to as the conditions of demand. The first of these conditions is the tastes and habits of users. The second is the price, quality and availability of commodities that users consider to be substitutes for the product. These two together should account for the users' willingness to purchase. The third condition is the incomes, assets and access to credit of users, which account for the ability to purchase.

Where the conditions of demand are stable, the graphical function can be used to represent not only price and quantity *differences* within a given time-period, as above, but also price and quantity *changes* in successive time-periods. This is legitimate only if expectations about the future are stable. Almost invariably, under these conditions, higher prices are associated with lower quantities and so the demand function slopes downward from left to right.

A matter of great interest in analyzing the economic demand for water is users' responsiveness to price differences. With any given difference (or change) in price, is the response large or small in terms of quantity purchased? One measure of this concept of responsiveness is called the price elasticity of demand and is equal to the proportionate difference in quantity purchased divided by the proportionate difference in price paid. In applying the concept of the economic demand for water, it is necessary for users to face a volumetric price or cost in accessing their supply. For example, on the Palestinian West Bank water utilities such as the Jerusalem Water Undertaking meter their supplies to households, to industry and to urban services. Moreover, rural households and farmers, although not on a networked supply, still face costs varying with volume when they pump groundwater.

10.6 SUPPLY-SIDE LEAKAGE AND EVAPORATION

In Table 10.1 the categories of supply include in row 16 leakage and evaporation losses that occur between the point at which human society appropriates outstream water and the point of delivery of water to user properties. Curiously such losses are almost always treated in the current literature as a form of 'demand'. The result is that consultants' forecasts of growth in 'demand' include (as unaccounted-for-water) supply-side leakage and evaporation. Furthermore the reduction of these losses in abstraction, storage and distribution prior to the delivery of water to the user is said to be a form of 'demand management'. This does not seem helpful. A manufacturer of refined sugar, when considering losses from output because of pilfering, or contamination while in the warehouse or destruction in a road or rail accident en route to the supermarket, would never regard this as a demand for sugar, a bizarre act of use by a consumer whom the sugar never reaches. The manufacturer would regard all of these as storage or distribution losses in the supply chain. So should it be with the supply of water.

There is a more general point here. Water resource management is now widely seen as principally a form of 'demand management'. To me such an approach seems nonsensical. This is especially the case when so much 'demand-management', as with reuse and the reduction of losses, turns out to be made up of supply-side initiatives. With outstream water we should always integrate in our thinking and our practice both demand-side *and* supply-side strategies.

10.7 CONCLUSIONS

This technical note concludes by illustrating some of the confusions that can arise when the concept of demand is deployed in the catch-all manner here criticized.

- To take demand to mean both use and consumption neglects the truth that a given volume used can be associated with wide variations in the volume consumed. This difference is at the core of much current debate in the field of irrigation engineering as the discussion lists show.
- To take demand to mean both use and need obscures the shortfall of use against need for more than half the world's population.
- To take demand to mean both use and the economic demand for water confuses the use of water with one of its determinants whenever costs borne by the user vary with the volume used.

If we professionals and policy-makers in water resources management cannot distinguish the use, consumption, need and economic demand for water from one another – who will? And if not now, when?

The material in this chapter originally appeared in: The demand for water: four interpretations. *Water International*, 29(1): 27–29, 2004.

11

The political economy of water abstraction charges

11.1 INTRODUCTION

Fresh water is a fundamental necessity for human society – for the individual citizen, for households and for all those sectors which contribute to the economy's output of goods and services. The principal sources of this water are threefold: first, directly from precipitation itself, most obviously in forestry and agriculture; secondly, the abstraction of surface water, that is, fresh water from rivers and lakes; thirdly, the abstraction of ground water, that is, from subterranean aquifers. Note that the desalination of sea water, at the global level, contributes only a small proportion of total supply (McDonald & Kay 1988: 8–34).

The focus of this article is the charges levied by government on individuals and institutions for the right to abstract surface and ground water. One objective is to provide a general economic analysis of the subject capable of international application to specific case studies. The second objective is to recommend the most appropriate basis for government abstraction tariffs.

The scope of abstraction charging includes water consumers abstracting directly for their own use, such as farmers, mining companies and manufacturers. One also includes water service companies that abstract not for their own use, but in order to provide the public water supply. In the Thames catchment in the UK, for example, the largest abstraction permissions provided by the Environment Agency are for six statutory water companies, plus a number of industrial users, such as Didcot power station, that are direct abstractors. Excluding precipitation itself, direct abstraction is also the main source of water for agriculture in the Thames Valley (Merrett 1997: 15–22).

The instream users of water are not addressed here, such as fishing, navigation and leisure, where no abstraction of water is required for such use. Nor does the paper deal with tradeable abstraction rights, such as those in Australia and Chile, because these constitute market transactions between private parties, with government merely setting a regulatory framework (Pigram et al. 1992, Hearne & Easter 1995).

11.2 ABSTRACTION CHARGES AND THE THEORY OF RENT

A useful analytic starting point is to establish that abstraction charges are a form of economic rent. The modern theory of rent was first developed in the early nineteenth century (Robinson & Eatwell 1974: 11–17). Its application was to British agriculture and the common situation where capitalist farmers used land owned by the aristocracy and the gentry, and paid rent for that right. At that time, agriculture contributed about one-third of Britain's total output. The classical school argued that the appropriation of land has the consequent effect of the creation of rent. Thus, Ricardo writes: 'Rent is that portion of the produce of the earth, which is paid to the landlord for the use of the original and indestructible powers of the soil' (Ricardo 1821: 67).

It is true that we may now doubt that any power of the soil is indestructible. Nevertheless, since Ricardo's time, economists have used rent theory whenever they are dealing with a natural resource of economic value, which can be privately appropriated, and which is in restricted supply. Clearly this makes rent theory applicable to natural supplies of fresh water. These flows are of economic value. Rights in their abstraction can be privately appropriated. Lastly, surface and ground water is in restricted supply in any given year. Winpenny (1994: 54–67) provides examples of rental markets for water in India and the USA.

However, the supposed inelasticity of raw water supply deserves closer scrutiny. Within a catchment, short-term elasticity is perversely high, where substantial stocks are held in a catchment's reservoirs or where aquifer stocks are high and ground water pumping capacity is not fully utilized. But it is the elasticity of the planning supply function of raw water over the medium- to long-term which is at issue here. It is certainly true that total rainfall in the catchment less evaporation sets a hydrological constraint on long-term catchment abstraction, as Dubourg suggests (1993: 3). But the recycling and reuse of water weakens this natural barrier whilst the desalination of sea water, and abstracted water imports from another catchment can smash it. The problem is that the wider one casts the net for additional supply, the greater are its unit costs. It is the exponential character of this planning function which maintains the truth of the statement that in the majority of the world's catchments with substantial populations, abstracted water is indeed in restricted supply (Turner & Dubourg 1993: 5).

Let us now return to the classical theory of differential rent. This assumed that competition existed between farmers for access to land, and between landowners in supplying land. In such a situation, it is theorized that land rent on the least fertile tract of land worth cultivating would be zero, and here the farmer would earn the going rate of profit on capital for the the private sectors of the economy as a whole. Land which was more fertile (or better located) would bring the owner a higher flow of rental receipts such that the farmer, after paying this higher rent, would still receive only the going rate of profit. Rent could therefore be seen as the transformation of the surplus profits of capital, driven by the competition for land of differential fertility.

Ricardo had only a limited interest in non-competitive markets in the supply of land for rent, albeit he did describe a case where monopoly rent, rather than differential rent, is paid (Merrett & Sharp 1991: 303). As a consequence, his theory is essentially demand-driven. This neglect of the supply-side determinants of land rent restricts the scope of differential rent theory in its application to abstraction charges levied by government. Such charges exist typically where an agency of the state licences abstraction to specified individuals and institutions, and where no property right exists to abstract ground and surface water without such a licence. Here, property rights on the supply side are assigned by a state monopoly.

Thus with government abstraction charges, analysis certainly requires an understanding of the effective demand for water by water service companies and direct abstractors. Differential rent theory is unquestionably valuable here. But analysis also needs a means to interpret the legislative practice of charge-setting by state institutions in different catchments, regions or countries.

11.3 A CHARGE-SETTING TAXONOMY

Having argued that abstraction charges are a form of economic rent, but that differential rent theory does not provide a basis to explain charge-setting by the state, the best way to proceed may be by developing a taxonomy of charge-setting principles. Such principles can then provide the basis for interpretive empirical work in specific catchments, as well as a starting point for policy review of existing legislative practice. Eight charge-setting principles are set out below and each is glossed in turn.

No charge. This is the lower limiting case. In the majority of the world's countries, government abstraction charges simply do not exist. In some cases the argument is that, since surface and ground water are a gift of Nature or of God, the state has no business in taxing it. This may explain the situation in Scotland, for example (Fowler 1995: 17–8). In other cases, what is lacking is the administrative capacity to levy the tax.

A revenue-maximizing charge. This is the upper limiting case. The long-term effective demand curve for water is probably cubic, reflecting the absolute necessity for the good, so in principle government could raise all its revenue requirements from this single tax (Merrett 1997: 53–8). In practice, of course, such a principle is never applied.

A market-clearing charge. In countries and regions which are arid, or where levels of water consumption are high in relation to effective rainfall, users may wish to gain access to more water than is available, at least in the absence of inter-catchment transfers. In this case, government may put in place a demand-management policy in which a general abstraction charge is applied which, although it is not revenue-maximizing, does just match the demand by abstractors to the annual flows available. No pure examples exist of this, the closest being the public auction of tickets for fixed time and flow in the centuries-old water market of Alicante in Spain (Winpenny 1994: 55–7). The adoption of demand-management policies, including water resource pricing of various types, is gaining ground on a global scale. For example, Schiffler and his colleagues at the German Development Institute have shown for the case of Jordan that supply-fix policies are under considerable pressure and that a demand-management philosophy is taking shape (Schiffler et al. 1994: 13–6). Politically, it is impossible at the present time in Jordan to tax water abstracted for use by farmers, but an abstraction tax on industry has been introduced there. Winpenny argues that one of the world's longest established and most successful water markets is in the Colorado-Big Thompson scheme, where the Northern Colorado Water Conservation District since the mid-fifties has used water pricing to transfer abstraction supplies from agricultural to urban uses, thereby raising allocative efficiency (1994: 59–61).

An environmental regulation charge. In this case one is considering a country which has a well-developed national policy for water resource management. The necessity for a regulator of the fresh water environment is accepted and abstraction charges are levied and hypothecated to finance the costs of regulation. ('Hypothecation' is a term widely used in the UK to refer to cases where government income from a defined tax is reserved for a specific expenditure category.) A clear example here is the Environment Agency for England and Wales. This institution's riverine management costs on current account are specifically recovered by charges made on licensed abstractors under a scheme of abstraction charges approved by the Secretary of State for the Environment (National Rivers Authority, 1993: 7–8, 1995). The Agency may revoke or revise licences but funds for compensation have to be found from its regional water resources budget (Rees & Williams 1993: 25–6).

An average total cost charge. In addition to regulation costs, the state may bear other costs in making fresh water resources available in a catchment. The abstraction of fresh water can require infrastructures provided by the state, sometimes on a considerable scale, as was the case in Early Mesopotamia. In such cases, a charge may be levied per cubic metre of water abstracted in order to defray in whole or in part the average total cost of these infrastructures. An example here is the abstraction charges levied on farmers in the Goulburn-Murray Irrigation District of Victoria in Australia (Pigram et al. 1992: 91–120). Another cost is the compensation it may be necessary to pay to existing holders of abstraction rights, when a catchment authority wishes to modify or rescind such rights in order to achieve the objectives of its catchment management plan. An average total cost charge is equal to total financial costs of regulation, infrastructural investment, compensation payments and any other miscellaneous items, divided by the total volume of water abstracted during the year. Its introduction would have many opponents amongst water users and water suppliers, despite the clear feasibility of such a charge being fiscally neutral. Moreover, in the opening up of new areas for irrigation as part of a regional development strategy, a charge on users reflecting the full cost of the water supplied to them would deal the *coup de grâce* to any such ambition. An interesting account of successful full recovery of deep tube-well installation and maintenance costs from farmers through abstraction charging is the Barind Multipurpose Development Authority's experience in Bangladesh since 1985 (Zaman 1996).

A marginal cost charge. Some writers working within the orthodox paradigm, such as Dubourg (1995), have considered the appropriateness of an abstraction charge which reflects marginal cost. There are two serious problems with such proposals. First, it has to be said that catchment authorities frequently have little or no interest in marginal additions to abstraction capacity. Marginal change is not the point, it is non-marginal changes in capacity that we need to address, as Dubourg himself points out (1995: 8). This moves the discussion, for example, to average cost per cubic metre of additional water abstracted and to the traditional utility price analysis dilemma in which large surpluses are generated when average additional cost exceeds the average cost of existing capacity and large deficits in the reverse case. Secondly, a catchment will have many infrastructural options in terms of scale and of type, capacity increasing, nature conservation, water quality improvement etc. So, many different average incremental costs exist, *some of which bring no increase in the volume abstracted or even bring reductions in that volume*. The average incremental cost per cubic metre of water in the no-change situation is infinite and with a cut in output is negative!

A Pigovian charge. Here, the principle is that government should set a charge which is differentiated according to the external costs imposed on society by each class of abstractors (Pigou 1932). K. William Kapp later formulated the parallel concept of 'social cost', covering 'all direct and indirect losses sustained by third persons or the general public as a result of unrestrained economic activities.' (Steppacher 1994: 440). An example of the kinds of damage done by over-abstraction is that of the Hueco Bolsón aquifer on the Mexican–US border: over a period of 70 years the water table fell by 25 metres, resulting in increased pumping costs, subsidence, and contamination by increased flows of saline and polluted waters into the fresh water source (McDonald & Kay 1988: 51–3). The second source of external costs is the pollution of abstracted water by use and the subsequent discharge of that water back into surface and ground water sources, at outfalls and even through injection wells. But this second type of externality does not, strictly speaking, arise from abstraction but from consumption. As a general rule, whilst abstraction charges may include a component in recognition of the polluter-pays principle with respect to over-abstraction and, sometimes, pollution by use, the *differentiation* of the charge on the basis of the monetary evaluation of the external costs never takes place, because of the extreme difficulty of measuring them. Indeed, political economy denies even that such measurement has any meaning. As Söderbaum (1994: 197) says:

> the idea of a 'correct level of pollution control' connected with some level of economic activity where marginal net private benefits intersect with marginal external cost are questioned, as is a 'correct level' of (Pigovian) taxes which leads us to the 'optimal level of externality'. While there is a monetary aspect in damage to ecosystems and human health from pollution, the institutionalist questions the reduction of damages and negative impacts to money and also the idea that theory can single out correct prices and correct charges as part of societal valuation.

Kraemer (1995: 231–2) refers to the overwhelming problems of applying a Pigovian tax in his introduction to the development of German abstraction charging between 1988 and the present.

An incentives charge. Incentive charging in a catchment or region can be defined as the use of a water tariff – a table of fixed charges – to give price signals to abstractors which reinforce water resource management based on environmental standards and regulatory controls. The best developed system of incentive charging is probably that of Germany. In contrast, sources within the Environment Agency suggest that abstraction charges in England and Wales are so low, they are incapable of significantly influencing abstractor behaviour.

An incentive tariff can include the following components:

- A licence fee to have a new water abstraction installation approved by the regulatory authority. For example, Foster et al. (1993) in their comprehensive introduction to the hydrogeological, legal and administrative aspects of ground water licensing in Latin America indicate that installation fees are common and are hypothecated.

- A charge per unit volume. This may be invariable with total volume consumed. Alternatively, it may rise with the volume drawn off. The total charge payable may be based either on the licensed volume or the actual volume abstracted. As Schiffler et al. (1994: 15) point out, a significant drawback of a rising block tariff is that it hits hardest those abstractors requiring large volumes simply because of the size of the farm or factory. It should also be observed that abstraction charging is never imposed where the installation's capacity falls below a minimum level set by government. The effect of this is that direct abstraction by households to meet their needs for drinking, washing, cooking and sanitation are outside the regulatory system.
- A charge reduction for the quantity of water directly returned to surface waters after use. To take the example of Didcot power station again, this has a licence to abstract 142 million litres of water per day. The licence requires 50–66% of the water abstracted to be returned to the river, depending on flow conditions. Unit price is lower because of this 'non-consumptive' use. In contrast, spray irrigation provides virtually no return flow to river or aquifer and so is undiscounted.
- A charge which is greater for higher quality water. This is found in Germany, for water drawn from deep aquifers, where the average total cost per cubic metre abstracted is inevitably higher.
- A charge which varies with the seasons. The volumetric rate is higher in those months when effective demand is greater and higher too when precipitation is less.
- A charge which is greater for certain locations. These include upstream sources, because the length of the river exposed to abstraction impacts is greater; rivers, lakes and aquifers most threatened by past or present overdraft; and regions with lower effective rainfall.

11.4 ABSTRACTION CHARGES AND SUSTAINABLE CATCHMENT MANAGEMENT

In this section, the subject of abstraction charging is related to the debate on sustainability, beginning with the approach of Richard Dubourg (1993, 1995), whose contributions to the study of hydroeconomics are within the neoclassical framework and are widely influential within the UK. Dubourg suggests that aggregate sustainability is a situation where natural capital as well as non-natural capital are non-declining and specifically where 'critical capital' such as water, within the natural capital category, is non-declining. From these definitions he deduces impeccably that aggregate sustainability is consistent with catchment management policies in which abstraction of surface and ground water is equal to effective rainfall, that is, the entire hydrological flow known as runoff, defined by hydrologists as total precipitation in a catchment less evaporation (McDonald & Kay 1988: 8–19). It has to be said that the adoption of such a definition of sustainable abstraction is extraordinarily dangerous. Abstraction at a rate equal to runoff is certainly capable of leaving surface and ground water stocks unchanged over the course of a year – the critical capital stock is, indeed, maintained. But a rate of abstraction at this maximum rate would have the effect of capturing the entire river flow at some point or points in the catchment. In effect, Dubourg's definition ignores completely what in the US is termed the environmental need for water in river systems – all on the basis of three mathematical constraints!

In *Introduction to the economics of water resources* (Merrett 1997: 147–8) I have suggested an alternative, multi-dimensional approach. After defining the concept of a 'sustainable society', it is suggested that water resource planning in such a society has six principal fields of action:

- the protection of water's hydrocyclical capacity to renew its ground and surface water flows and stocks;
- the conservation of society's species and natural habitats in all their fresh and salt water environments;
- the husbandry of water in its supply and use;
- the supply of fresh water sufficient to meet the biological, cultural and economic needs of society's human populations;
- the purification of water from domestic, agricultural and industrial effluents;
- the drainage of storm water, and the protection of rural and urban communities against floods.

Abstraction charges can contribute positively to the first three of these fields and below it is shown how this can be done, using what I shall call *full incentive charging*.

The preparation of a tariff scheme should take as its starting point the regulatory controls put in place by the catchment authority in carrying out its responsibilities. In the development of these controls, there is an important role for social cost effectiveness analysis and social cost benefit analysis. These techniques have a place, alongside environmental impact assessment, in comparing alternative regulatory options. The design of environmental regulation is, then, in part an economic process as well as an ecological one.

The tariff régime introduced to any specific catchment (or region) will be contingent on its physical geography, its habitats and species, its human settlement patterns, its economy and the power relationships which hold between various social and economic groups. So what we require are criteria for tariff design which can be applied in the appropriate way to any single area with all its unique characteristics. I propose three such criteria for full incentive charging.

First, the annual income from abstraction charges should be hypothecated to the environmental regulator such that, when added to the income received from discharge fees, fishing licences, navigation permits etc., abstraction fee income is sufficient to cover all the state's capital and current account expenditures on environmental regulation, water infrastructures, research and database development, compensation payments, etc. The justification for this criterion is that water users, who in the final analysis are certain to bear the charge, should pay the full financial cost of water provision. In providing the regulator with hypothecated income, it is likely to increase their relative power as an agency of government. As Kraemer (1995: 238) writes of Germany since 1988: 'On the whole, the water resource taxes contributed to capacity building within the water management administration in the German Länder and thus partly overcame the implementation deficit in water resource management'. It will also provide a budget to finance the legal costs of modifying or terminating abstraction licences which threaten the hydrocycle or undermine nature conservation. Higher abstraction charges will also give a price signal to water companies and direct abstractors to reduce their storage and distribution leakage between the points of abstraction and the user's gate.

Secondly, where the full cost abstraction tariff still leaves an excess of demand for abstracted water over its licensed supply, the charge should be raised so that market clearing takes place.

Thirdly, specification of the components of the abstraction charge should provide incentives for abstraction behaviour that is economically efficient and that avoids environmental degradation. Price per unit volume should be invariant with total volume abstracted, unless there are countervailing economic or environmental arguments. Here, one accepts the argument in section three above by Schiffler et al. against a rising block tariff. Price should be discounted where abstractors recycle their off-take to surface or ground water sources. (Discharge fees should be used to handle the water quality aspects of recycled water.) Charges should be higher for abstracted water of higher quality. Seasonal variations in effective rainfall and economic demand should be dealt with through the licensed volume provisions laid down by the regulator and by the market clearing criterion. Charges should be higher for upstream sources and for abstraction in locations where species and habitats are more threatened by abstraction. Charges should be higher in parts of the catchment (or region) where water-related infrastructural costs borne by the state are higher and should be higher to reflect the greater infrastructural costs generated to meet peak demands.

11.5 THE IMPACT ON USERS

What can one say about the impact of full incentive charging on the purchase of abstracted water? Here the analysis is best conducted in terms of the big battalions – water companies, industry and agriculture.

Water companies with a statutory responsibility to provide the public water supply would readily acknowledge that the current abstraction charges they face are passed on to final consumers. In any case, such charges make up only a small part of total annual costs. The annual expenditures of a water company are dominated by the cost of capital, staff wages and salaries, electric power and materials such as chlorine (Merrett 1997: 128–34). So even a two- or five-fold increase in the abstraction charge per cubic metre of water would have little effect on the price to consumers of the public water supply. The principal response to a change in the charge is likely to be a review by the company of the cost effectiveness of reducing the leakage rate in its production and distribution systems. The response of consumers and of the water company would, nevertheless, reduce the quantity the company seeks to abstract. Unfortunately, no empirically based estimation of the elasticity of demand for water abstracted by statutory companies exists, as far as I am aware. Of course, in countries such as the UK where most households pay a fixed charge for water consumed, not a price per unit volume, there would be no demand response whatsoever.

Now let us consider the case of direct abstraction by industry, such as in power generation, mining, construction, manufacturing and the services sectors. As a general rule one can say that where industrial companies use only small quantities of water, their reduction of quantity purchased in response to full incentive pricing is likely to be negligible. Rees et al. (1993: 17, 40) have suggested that the costs of water to UK manufacturers are usually a maximum of 0.2% of total costs and that industry exhibits a very low level of knowledge of water's volumetric price even when purchased from water companies. However, where industrial processes require large volumes of water, there may be a strong interest in water reuse within the company, with a consequentially large fall in the quantity which needs to be abstracted. Reuse also has the advantage of cutting firms' fees for the discharge of industrial effluent.

Throughout the world, water is also directly abstracted by individual farmers or by organizations acting for them. Water for irrigated agriculture constitutes more than 90% of total consumption in a large group of low-income countries (Kinnersley 1994: 181). In these cases, full incentive charging may have a powerful impact on farmers' budgets. Where farmers' incomes are low and full incentive charging would make water unaffordable for them, and where government or a delegated agency wishes to continue to subsidize the agricultural sector, such subsidies should be transparent. Where full incentive charging *is* applied, it is likely to stimulate husbandry in the application of water as well as the production of crops with high water productivities, as Schiffler et al. (1994) and Gleick et al. (1995) have shown for Jordan and California respectively.

11.6 FINAL REMARKS

I wish to end this paper by suggesting the institutional framework which would be most appropriate for full incentive charging. These ideas have been strongly influenced by Karin Kemper's *The cost of free water* (1996). The basic approach is a negotiation model in which there exists a public sector catchment agency that, through a negotiating forum, develops its policies with the advice of water companies, direct abstractors, environmental organizations and water user associations representing the domestic sector, irrigated agriculture, mining and manufacturing, etc. The original prototype for such negotiation models is the French water parliaments (Tuddenham 1995). Central government retains the statutory right to determine which public and private institutions may enjoy the right to abstract water, on what scale, in what locations, at what time of year, at what price – but it delegates such rights to the catchment agency.

The agency, with the assistance of its partners in the negotiating forum and with a full understanding of existing customary rights, then assigns formal abstraction rights on a time-limited basis to specific abstractors. The time limit would be 10 years, let us say, rolled over each year except where the agency allows the licence to expire. Where these 10-year rights need to be modified or rescinded for hydrological, environmental or economic reasons, compensation would be payable to the abstractor so affected. Rules would also exist to address Third Party impacts. Abstraction rights assigned to abstractors could be freely traded between abstractors provided the agency had approved such transfers in the light of their environmental impacts.

Full incentive charging would be the basis of the price paid by abstractors for their water and has the objectives of underpinning environmental regulation with a hypothecated income source, of requiring abstractors (and therefore final users) to cover the full costs of regulation and state infrastructural provision, of bringing the quantity of water demanded by abstractors into line with regulatory limits, and of giving price signals which promote both allocative efficiency as well as abstraction practices which avoid damage to riverine eco-systems. No charge would be made for abstractions below a minimum scale; the transaction costs of such charges would be high relative to the volumes abstracted.

Abstraction infrastructures would be constantly reviewed. Where they require rehabilitation or new investment, this may be carried out either by the agency or the abstractors themselves. In the former case, the financial costs would be reflected in the agency's charges, as already indicated. Arrangements would be made to monitor abstraction with respect to its location, time and quantity, as well as to invoice abstractors, to collect the charges owed and to enforce all agreements. Such transaction costs would be included in the full costing tariff of abstraction charges. The creation of this institutional framework imposes social, economic and political costs on the parties concerned, that is, structural costs of change both real and financial in their nature. Therefore the negotiating forum may agree that it is sensible for full incentive charging to be phased in gradually.

One final issue is briefly considered here. It is argued above that regulatory control is the context within which abstraction charging is deployed. This may suggest that the environmental constraints are designed and specified and only thereafter is the incentive tariff formulated. In practice, work on both should proceed simultaneously and dialectically, because regulation and environmental pricing are interdependent activities within a single planning process.

The material in this chapter originally appeared in: The political economy of water abstraction charges. *Review of Political Economy*, 11(4), 1999.

12

Twelve theses on the cost and use of irrigation water

12.1 THESIS 1

In the field of water resource management a widely held belief exists that allocation stress is found in many parts of the world and is set to become more intense in the future both because of climate change as well as global growth in population and income. Allocation stress refers to access conflicts between the agricultural, domestic, industrial, urban service and environmental uses of outstream and instream water flows. If the belief that allocation stress will intensify is well founded then, because of the dominant role of irrigation water use at the global level, it is worthwhile exploring the possibility of slowing the growth of farmers' use of such water.

12.2 THESIS 2

To some degree the deceleration of irrigation water use can be achieved by raising the cost of water to farmers. The higher cost may lead them to reduce the cultivated area under irrigation, or to favour crops that use less irrigation water, or to apply water more productively, or some combination of these three actions.

12.3 THESIS 3

Raising the cost of water to farmers is unfavourable to them as a class. It is likely to reduce their gross margins, or require investment in water-saving technology, or reduce the market value of their holdings, or in some cases drive them out of business. Farmers can be expected to resist the new cost policy more-or-less forcefully.

12.4 THESIS 4

In all cases, today and since the dawn of irrigation in Mesopotamia, farmers' costs of irrigation water have taken one of four forms: own-supply cost, volumetric price, indirect fee or revenue-only fee. Hybrid types also can be found. Own-supply cost refers to any case where the farmer provides (at a cost) his own supply of irrigation water, such as in the case of building and maintaining his own well. Volumetric price refers to those cases where a farmer purchases irrigation water at a price per unit volume. Indirect fee indicates the many and varied cases where a farmer pays for his water under rules in which the cost he bears has an indirect relation with the volume used. Revenue-only fee indicates cases where a farmer pays for his water but where no relation exists between the payment and the volume of irrigation water used.

12.5 THESIS 5

Own-supply costs, certainly with respect to their prime cost component, vary with the volume of water used by the farmer, that is, the farming family's or agribusiness's field supply. This *binding* of cost and volume may be loosened by implicit subsidies, particularly cheap energy. Here the cost of irrigation water to farmers can be raised by reducing energy subsidies. It can also be increased by the introduction of abstraction charges.

12.6 THESIS 6

Volumetric price also binds cost and volume. But the price may fall short of the full-cost of the field supply. Indeed it may fall short of operation, maintenance and management costs. Here the cost of irrigation water to farmers can be raised by increasing tariffs. However, the introduction of volumetric pricing faces grave difficulties: flow meters require delivery through a pipe resulting in head loss that may reduce water availability at the farm level; and the cost of installing meters, monitoring them and collecting fees from smallholders may be prohibitive.

12.7 THESIS 7

Indirect fees are charges for irrigation water that have an indirect binding of cost and volume. They include charges that vary with the farmer's irrigable area, or for a water turn of given time duration, or with crop composition where 'wet crops' incur heavier charges than 'dry crops'. Indirect fees are often accompanied by a quota regime. Here again the cost of irrigation water to farmers can be increased by raising tariffs.

12.8 THESIS 8

Revenue-only fees are those that do not meet the binding principle but merely act to raise income for an irrigation authority. They can be used to reduce irrigation water use only by driving farmers out of business.

12.9 THESIS 9

Where the binding principle operates (see Thesis 5) the possibility exists that an increase in irrigation water costs may lead to a deceleration in the growth of use (or a fall in use) in comparison with the no-cost-increase alternative.

12.10 THESIS 10

A policy to decelerate significantly irrigation water use by raising its cost to farmers requires that a minimum absolute value of the elasticity of demand exists, for example a value of 0.5. Such a minimum value requires some combination of the following conditions:

- The binding principle operates.
- The farmer exercises control over the field supply volume.
- Irrigation water costs are a visible component of prime costs to the farmer.
- The farmer understands that a proportion of the water applied with the current technology (and at specific stages of the crop year) enjoys a low return in terms of crop yield.
- The farmer understands that new techniques are available that will improve the productivity of water applied, such as its gross margin efficiency.
- The farmer understands that reducing the volume of irrigation water used does not materially increase the risks s/he faces.
- The farmer understands that the losses of non-crop returns from irrigation water reduction are not large.
- Crops are not 'feed crops', that is, they are not cultivated for the farming families' subsistence, or for feeding to their livestock, nor are crops grown by agribusiness as an input to its own manufacturing processes.
- Under conditions of general (even if modest) inflation, the cost of water to the farmer must outstrip the wholesale price of the crop.

12.11 THESIS 11

At the basin level the good sense of water management policies tackling allocation stress by reducing abstraction for irrigation holds only if the water thereby released actually becomes available for domestic, industrial, urban service of environmental uses.

12.12 THESIS 12

The reduction of the base supply to the irrigation sector reduces the return flows of water to farmland, river, lake and aquifer. In some cases these lost returns consequent on diminished abstraction can wipe out the advantages of reallocation.

The material in this chapter originally appeared in: Twelve theses on the cost and use of irrigation water. *Irrigation and Drainage*, 51(3): 265–268, 2002.

13

Behavioural studies of the domestic demand for water services in Africa

> *... there is no such entity as a distinctive type of 'structural explanation' in the social sciences; all explanations will involve at least implicit reference both to the purposive, reasoning behaviour of agents and to its intersection with constraining and enabling features of the social and material contexts of that behaviour.*
>
> Anthony Giddens, *The Constitution of Society*, 1984

13.1 A METHODOLOGICAL REVOLUTION

The purpose of this paper is to contribute to the modern history of hydroeconomics. From the mid-1980s a very small group of university academics, and professionals from the international banks, launched a methodological revolution in the study of domestic water and waste water services in the low-income countries. This revolution resulted in a permanent transformation of the way the water resource professions approach the household sector there. Here I wish critically to review this literature in order to strengthen future research. The paper does not cover techniques for forecasting future price and volume outcomes based on willingness-to-pay field work; it is limited to the study of existing behaviour. Willingness-to-pay surveys are the subject of a future article.

The three principal authors of this change were Dale Whittington from both the University of North Carolina at Chapel Hill and the World Bank, Xinming Mu of the Asian Development Bank, and Donald T. Lauria, also from the University of North Carolina. Other authors were Briscoe, Roche and Wright of the World Bank, Choe, Hughes, Okun and Swarna all at Chapel Hill, Okorafor and Okore at the University

Table 13.1 Seven behavioural studies of African domestic water services.

Authors	Case study location	Year of field work	Year of publication
Whittington, Lauria, Okun, Mu	Ukunda, Kenya	1986	1989
Whittington, Mu, Roche	Ukunda, Kenya	1986	1990
Mu, Whittington, Briscoe	Ukunda, Kenya	1986	1990
Whittington, Lauria, Mu	Onitsha, Nigeria	1987	1991
Whittington, Okorafor, Okore, McPhail	Nsukka, Nigeria	1989	1990
Whittington, Smith, Okorafor, Okore, Liu, McPhail	Nsukka, Nigeria	1989	1992
Whittington, Lauria, Choe, Hughes, Swarna, Wright	Kumasi, Ghana	1989	1993

Source: see references.

of Nigeria, Liu and Smith at North Carolina State University, and McPhail at Johns Hopkins University. The key publications are listed in Table 13.1. As can be seen, the field work was carried out in the period 1986–9 in Africa, and the papers were published in the period 1989–93. Four areas were researched: Ukunda in Kenya, Nsukka district in Nigeria, Onitsha in Nigeria and Kumasi in Ghana. The focus in these first three areas is water services; in the fourth it is waste water services.

Whittington and his colleagues show a strong interest in policy, suggesting that the low-income countries in general have suffered from repeated failures in programmes devoted to domestic water and sanitation needs. 'There are simply too many leaking taps, abandoned water systems, and defunct village water committees for anyone to be sanguine about the current rate of progress' (Mu et al. 1990: 521).

It is forcefully argued that these weaknesses flow from traditional approaches that are 'out of touch with present demographic and financial realities' (Whittington et al. 1993: 733) in which 'designs for new systems are generally made and projects constructed with little understanding of household water demand behaviour' (Whittington et al. 1991: 179). Master plans of sewers and treatment plants as well as engineer-dominated supply-side philosophies are criticized for their neglect of the demands of beneficiaries. The view, ascribed to van Damme & White (1984) that design specifications should not require households to pay more than 3–5% of their income on water and waste water services repeatedly comes in for attack. In fact the setting of water charges within the 3–5% bracket goes back at least as far as 1975 (World Bank 1975).

As a consequence, the authors argue that the way forward is to undertake detailed case studies of families' actual water use behaviour and their observed ability and willingness to pay for water services. This information can provide a well-grounded basis for estimating the uptake of new programmes (Whittington et al. 1989, 1990a). In their first Nigerian study, Whittington et al. conclude (1991: 194): 'Most people in Onitsha are already paying high prices for water from the vending system for service which is inferior to that which could be provided by a well-run piped distribution system'; and the local water and sewerage authority should view itself 'as a regulated utility, not as an agency providing a social service'.

In what follows, I consider first the principal insights of these valuable contributions by Whittington et al. and their colleagues and then develop a critique of some aspects of their work. Two tables that may be useful for future studies are proposed: an Attribute Matrix and a check-list of some information requirements for household surveys.

13.2 MARKET NETWORKS FOR WATER

The water services in any region can be represented as a three-layer cake. The bottom layer is the forms of access to fresh water; the middle layer is the modes of distribution of the volumes initially generated; and the top layer is the pattern of uses of water. Taking all categories of use together, the total access volume less evaporation and leakage provides a net supply mathematically identical to the total volume distributed which, after again deducting losses, is identical to the total volume used. These identities provide the type of internal accounting checks I have employed in developing the concept of the hydrosocial balance (Merrett 1997, 1999, 2002).

The *first* strength of the water demand school that has been introduced above is its demonstration of how socially complex these networks of access and distribution can be. The 1986 study of the village of Ukunda with its population of 5000 demonstrates this well. Some households queue to collect their water directly from local wells and hand-pumps, performing all three roles of abstractor, distributor and user. In other instances a very small number of households have piped water to their homes sourced by the local water utility. In the third case kiosk owners receive a piped supply which is then sold directly to households who carry it back to their dwellings in 20-litre jerrycans. In the fourth case kiosk owners sell on the water to retailing vendors who themselves then distribute it to households' front doors, using carts or bicycles to carry the jerrycans (Whittington et al. 1989). The water demand school's work on the cities of Kumasi and Onitsha, with their populations of 600,000 and 700,000, respectively, showed each had its own distinctive access-distribution-use system. In these two urban cases, water tankers played a major role. In demonstrating these social complexities, Whittington and his colleagues also contribute greatly to our understanding of the economics of the supply processes.

The *second* strength of these research studies is their detailed examination of the dynamic and competitive markets in water that exist in many of Africa's villages and towns. In the Ukunda and Onitsha case studies it was possible to draw a chart of the access and distribution network to households, to collect data on the unit price for each branch of the network, to estimate the volumetric flows of abstraction and distribution and therefore to calculate the turnover along that branch.

Table 13.2 illustrates the case of Onitsha. With respect to households there are 11 relevant access/distribution branches. The variation of volumetric flow between branches is high. For example, the lowest dry season branch flow (shallow well collection by doorstep vendors) is a mere 10,000 imperial gallons per day (2200 litres/day), only 0.3% of the highest flow (private borehole sales to tanker trucks) at 3 million g/d (660,000 l/d).

Unit prices are also marked by their high variance. Shallow well collection incurs no monetary charge, although the collection time (the time spent queuing, drawing water and the journey to and from the well) can be lengthy and exhausting. Even limiting our scope to priced supplies, there is still a marked divergence. For example, the unit price of the public piped supply at 0.003 naira/gallon (0.0007 n/l) is only 2.4% of the price charged by doorstep vendors to households at 0.127 n/g (0.028 n/l).

Table 13.2 Water and money flows in Onitsha – 1987 dry season.

Access/distribution branch	Water volume (million gallons per day)[1]	Unit price (naira/gallon)[2]	Turnover (naira)
Household collection from private boreholes	0.30	0.027	8000
Private borehole sales to tanker trucks	3.00	0.003	10000
Truck sales to households	1.00	0.020	20000
Truck sales to shops	1.70	0.020	34000
Shop sales to households	1.55	0.050	78000
Shop sales to doorstep vendors	0.05	0.040	2000
Private borehole sales to doorstep vendors	0.05	0.020	1000
Shallow well collection by doorstep vendors	0.01	0.000	0
Doorstep vendor sales to households	0.11	0.127	14000
Shallow well collection by households	0.30	0.000	0
Public piped supply to households	1.50	0.003	5000
Total or weighted average[3]	9.57	0.018	172000

Source: Whittington et al. 1991, Figure 1.
Notes: [1]One imperial gallon = 4.546 litres.
[2]In 1987 the rate of exchange of the US$ to the naira was 1:4.3.
[3]Tables of this branch type inherently embody double-counting. The total supply to households is 4.76 million g/d and total turnover from sales to households is 125,000 naira.

The most extraordinary aspect of the monetary data, for those unfamiliar with African water markets, is that the proportion of water supplied to households by the public sector Anambra State Water Corporation in the 1987 dry season was 32% of the total but its share of total turnover was only 4%.

In Onitsha, Whittington et al. also note the powerful effects on the market of the changing seasons, dry and wet. The main difference is that in the rainy season households manage to capture 2.3 million g/d (10.5 million l/d) of rainwater with the effect that their purchases from the water vending system fall to about half the dry season total. Rainy season prices are lower, of course, with the downward shift in the demand schedule. Seasonal variation in the structure of the market is also observed in Ukunda, where rainy season prices are half those of the dry season, as well as Nsukka district.

The *third* strength of the water demand school is their success in estimating the proportion of domestic income absorbed by water purchases. In Ukunda annual per capita income is put at US$350 in 1986, annual domestic expenditure on water per head is about US$30, giving a per capita outlay/income ratio of some nine per cent. 'Those who use vendors enjoy a high level of service; good quality water is delivered to their doorstep on demand. Although vending does not provide a level of service comparable to house connections from a well-run piped distribution system, it is far superior to that available in most rural communities in Kenya. This high level of service is, however, expensive' (Whittington et al. 1989: 164).

In Onitsha in 1987, average annual household income was about US$1600, although about 25% of households received less than US$600. Average household size was six to seven people. Whittington et al. do not present a single value for the outlay/income ratio over the whole year and for all income classes, but their published information allows this to be estimated at about 12% in the dry season, 5% in the rainy season with an annual average of about 8% (Whittington et al. 1991). What is also clear is that the poor pay a far higher proportion of their income on water: 18% in the dry season and 8% in the rainy season. In this case the income elasticity of demand for water services is positive but less than one. 'In the past, it has been commonly assumed that households could only afford to pay 3–5% of their income for improved water services ...' (*ibid.* 189). But, the data from Ukunda and Onitsha, as well as studies by Linn (1983) in Ethiopia, by Fass (1988) in Haiti, and by Cairncross and Kinnear in Sudan (1991, 1992) refute this assumption completely.

For Nsukka district and Kumasi the treatment of the outlay/income ratio is much slimmer. In the former case, annual water purchases from vendors are probably between 6% and 10% of household income, but the estimate is judged to be 'highly speculative' (Whittington et al. 1990, 1903). In the latter case, average household expenditure on water and sanitation combined is about 3% of average household income but about 10% of households spend more than 8%.

Abraham Maslow's lexicographic approach to a five-level hierarchy in human need satisfaction receives confirmation here: life-or-death need for the service, high service cost and low household income combine to give a staggering proportion of income devoted to water and waste water services (Maslow 1954).

This conveniently brings us to the *fourth* strength of these authors' research: an innovative application of the access/distribution chart (see above) specifically to sanitation (Whittington et al. 1993). Kumasi's 600,000 persons are estimated to fall into five groups: those using public latrines (38%), those using bucket latrines (25%), those with access to a WC emptying into a septic tank (25%), those using traditional pit latrines (7%) and those who urinate and defecate on open ground (5%).

The monthly total of this feculent mass is some 25,000 m^3, which goes to open street drains, streams, neighbourhood dumps and landfill. Unit costs are estimated for the use of alternative sanitary facilities as well as the flow of funds to sanitation actors, including rent payments to the local Committees for the Defence of the Revolution.

The authors conclude (*ibid.* 745):

> Our survey of sanitation conditions in Kumasi revealed an appalling and, from a public health perspective, dangerous situation. Households are currently generating about 25,000 m^3 of human waste per month (including flush water for WCs), but only about 10% of it is removed from the city. The rest, 90%, is left in the urban environment until it decomposes, is carried away by small streams or drainage ditches, or dries and becomes airborne.

13.3 THE USES OF WATER

Above I have reviewed the principal strengths of the water demand school in these studies published in 1989–93. What follows is a critique of some aspects of their methodology, beginning with the treatment of water consumption.

In terms of economic theory the demand for a product (or service) can be defined as the relationship, at a given time and within a defined market, between price per unit of the product and the quantity in each time period that consumers are estimated to be willing to purchase at each price. Demand is conventionally represented graphically, with price on the vertical axis and quantity on the horizontal axis, showing the difference in quantity purchased at each price. The cultural and economic context of any demand function is referred to as the conditions of demand. The first of these conditions is the tastes and habits of consumers, that is, the nature of need for the commodity. The second is the price, quality and availability of commodities that consumers consider to be substitutes for the product. These two together should account for the consumers' willingness to purchase. The third condition is the incomes, assets and access to credit of consumers, which account for the ability to purchase.

With respect to the first of these 'conditions of demand', the nature of the need for water is complex because, even when we limit our interest to the domestic sector, the uses of water are extraordinarily diverse. This accounts for the huge variance in household water consumption in litres per head per day (lhd) within and between societies. At the most basic level, there is a biological drive, a physiological necessity, because water is the basic solvent necessary for all the body's functions, both at the biochemical and the cellular level. This suggests that a lexicographic classification of the domestic uses of water is appropriate (Earl 1995: 41–2). This could be:

- drinking;
- cooking;
- washing ourselves;
- disposal of urine and faeces;
- cleaning and washing in the house;
- garden watering;
- cleaning and washing of a car and filling of a swimming pool.

In general in the low-income countries, the first step in understanding use is to distinguish instream use (such as bathing or washing clothes in a pond) from outstream use. The greater the availability and convenience of instream supply, the less will be the demand for outstream water in the home. The second step is to exclude from household use the water that is applied for the irrigation of crops and the watering and cleaning of livestock. At the centre of an urban area this is unlikely often to be an issue, but on the urban periphery or in villages it can give misleading totals for the use types listed above. The third step is to take a view on the total volume of water used by a household per time period (allowance made for seasonal change) and its proximate distribution between the lexicographic uses. The fourth step is to examine what relation exists between the multiple uses of water and its multiple sourcing; for example high-priced, high-quality water may be reserved for drinking and cooking whereas turbid or malodorous water may be limited to waste disposal or garden watering. The fifth (and not the least important) step is to determine which members of the family use the water. Commonly it is the women who not only fetch water but who also wash the children, cook, clean the house and wash the clothes. It is they who best understand the complex choices to be made about water supply and water use and it is they who provide the best source of information on household behaviour. The sixth (and here the final) step is to examine household water use efficiency, the scale of reuse and how water is stored in order to balance the temporal fluctuations of supply and demand. Use efficiency is likely to correlate directly with unit cost, particularly when user and collector are the same person.

The ellipsis of the water demand school is that no account whatsoever is given in the published papers referred to in Table 13.1 of how African households in general, or their survey respondents in particular, *use* water. Simply put, to understand the demand for water requires, as a necessary condition, that the researcher knows how it is used. Whittington and his colleagues fail to meet this condition.

13.4 OBJECTS OR SUBJECTS?

The water demand school is open to a second methodological attack, related to but separate from their neglect of water consumption. I shall call this their naturalistic bias. Given their policy interests, Whittington et al. need to understand why it is that households make specific decisions on the sources from which they choose to gain access to water, the volumes they purchase, and (as above) how water is used. A useful technique in such explanation would be an attribute matrix of the type set out in Table 13.3, but such a matrix is never employed. Their approach is to collect a large dataset that covers, for each household, social variables (such as the number of adults and children in the household), economic variables (such as the household's total monthly cash income) and water practice variables (such as the volume of water accessed from each alternative source). These classificatory and quantitative datasets are then used as inputs to econometric modelling. Table 13.4 provides a possible template for such information requirements, including coverage of the uses of water.

An alternative approach, that would itself have provided the basis for *designing* an attribute matrix and *designing* an econometric model, would have been set within the framework of academics such as Veblen (1919) and Simon (1959, 1992). Peter Earl (1995) provides an excellent review of this paradigm. This approach would regard water access and use decisions by households as a process of problem-solving during which their views of the world and their wants may undergo considerable evolution, yet where at the date of any cross-section survey domestic behaviour takes the form of routinised, habitual action (Hodgson 1994). The door to the household's mind-set and actions in this regard would have been through semi-structured interviews with its members in which Africans speak for themselves, they voice the constraints they may face, they articulate their own accounts of what their routines are and why they satisfy a family's needs (and change with the seasons) rather than alternative habitual actions. Household members, in this case, are regarded as active, intelligent, resourceful, purposive, reflexive subjects, not merely as objects exhibiting quantifiable behavioural patterns.

A quotation from Earl's critique of orthodox economics applies with full force to the water demand school (1995: 61): 'Neoclassical economists working on consumer behaviour have preferred to steer well clear of this approach ... Instead, they have tended to confine themselves to studies of choice in relation to characteristics that are easy to measure in supposedly objective terms.'

Anthony Giddens in *The Constitution of Society* (1984) captures the method of authors like Whittington to submit it to a higher level of ontological critique. He suggests that social scientists who use frameworks that treat human subjects as objects are committing a naturalistic error. They turn their back on 'the active, reflexive character of human conduct'. Giddens also rejects orthodox social theory's vision of 'human behaviour as the result of forces that actors neither control nor comprehend'. He defines the naturalistic standpoint as that which looks to the natural sciences, including biology, as a model for social science.

Table 13.3 An attribute matrix for household water services.

Source	Attribute								
	Price per litre	Payment arrangements[1]	Collection and queuing time per trip	Odour	Turbidity	Other qualities	Pressure	Service times	Service reliability
Household collects rainwater									
Household collects from house well									
Household collects from other well									
Household collects from stand pipe or pump									
Purchase from doorstep vendor									
Purchase from doorstep tanker									
Purchase from shop or kiosk									
Piped supply to house									

[1] Such as a fixed monthly charge or volumetric price collected at the time of sale.

Table 13.4 Water and waste water services for households: information requirements.

The Survey Area
Map including residential locations and water sourcing points (wells, kiosks, etc.)
Number of resident persons
Number of resident households
Brief review of the survey area's economic base, its social character and its transport links to external areas of employment
Annual rainfall
Seasonal rainfall
Note: in cases of strong seasonal differences, data will be required separately for each season

Household and Water Use Information
Date/time/place of interview
Position in family of person(s) interviewed
Household size
Gender and age composition of household
Gender and age of persons collecting water
Education level of the household head
Occupation of all adults in household
Total household income per week (alternatively: total weekly expenditure)
Total weekly expenditure on water services
Address/location within the survey area
Ownership of water-using fixtures: bath, shower, WC, sink, dish-washer, washing machine, garden-watering items
Volume of abstracted water used in litres per head per day (lhd)
Composition of use (%): cooking and drinking; bathing, showers and personal washing; washing of clothes, house, vehicles, etc.; sanitation; garden; other
Types of use that are specific to types of source (if any)
Information on water reuse: waste water source to type of second use; no. of cycles, volume of reuse as multiple of initial use
Water storage: type of storage by volume

Attribute Matrix for Water Services
See Table 13.3

Waste Water
Type of service: water closet with piped connection to sewer; own latrine; communal latrine; septic tank; open areas
Type of treatment at point of service
Means of removal of urine and faeces: flushing to sewer or septic tank; collection from latrine; direct to lake or river; other
Charges for: service, collection, treatment and disposal
Total weekly expenditure on waste water services

Other Information
Owners/managers of water sources and waste water services
Action taken by owners/managers in event of non-payment
Water and waste water tariffs and billing arrangements established by source owners/managers
Type of dwelling, including storey height in case of flats
Number of rooms
Total weekly expenditure on power (gas, electricity, other fuels).

13.5 THE DISCRETE CHOICE MODEL

We have already seen that in Ukunda and Onitsha a considerable volume of quantitative data had been collected from households and market institutions. Modelling household behaviour could have been carried out by designing the equations in the light of social theoretical, semi-structured household interviews – a subject-oriented approach. Whittington and his collaborators' neglect of the reflexive character of household members did not permit a design of the model in this way. In practice they rejected multiple regression equations and opted for a conditional multinomial logit model where the dependent variable is the probability that any specific water source of several available is chosen by the household. This probability is expressed as a function of the price of water from each source, its collection time, its taste, household income, education of the household head and the number of women in the household. The model was used in the Ukunda study where water's price and collection time were both highly significant (Mu et al. 1990).

On first sight, the approach seems useful. But on careful consideration it appears to rely on the assumption that households choose only *one* of the alternatives available. This does not square with the fact of use of multiple sources by each household, amply illustrated in the Nsukka district. Nor does it address the volume of water taken from one or other source, the traditional dependent variable in modelling household water demand. Nor (as we saw above) is it based on an understanding of water use, the dog that does not bark in Whittington's world.

The rationale for the orientation to a discrete choice model is clearly policy-based:

> "... in an African village the first element in a water demand model must be a description of the likelihood, under different conditions, that a consumer would choose to use an improved source rather than continue to use the existing traditional source(s)" (*ibid.* 521).

The decision to use the discrete choice model, containing as it does no quantity-price relation, also has the regrettable outcome that Mu et al. cannot pursue mainstream interests such as the nature of the demand function for water (linear, exponential, constant elasticity, or cubic?) and the values of the demand elasticities.

The principal challenge offered by multi-sourced water, where cash price per litre can be traded off against collection time per litre, is how to incorporate subjective perspectives into a modelling framework that seeks to explain price and volume data and recognises explicitly:

- the multi-source environment;
- the simultaneous use by households of more than one source;
- water quality differences;
- the proportion of household members who are adult women;
- the female labour force participation rate in the survey area;
- the average hourly income for women from wage-labour or self-employment in the local market economy.

There is a further difficulty with the water demand school's approach and this can be illustrated by a quotation from their first publication (Whittington et al. 1989: 165):

> information on the water vending system can serve as a useful indicator of a community's ability and willingness to pay for a piped distribution system.

> The fact that yard taps do not already exist throughout Ukunda indicates an inability on the part of the community or water authority to mobilize resources, not an inability or unwillingness of the population to pay for the cost of the improved service.

But such a conclusion is hazardous. Whittington and his collaborators here fail to recognise the force of two key economic characteristics of the system they describe:

- The tariff for water in every case is volumetric, whether the cost is in collection time or in Kenyan shillings.
- The penalty for unwillingness or inability to meet these costs is an immediate cut-off in supply.

Only in the most unlikely case that a piped system is metered and where billing sanctions are draconian could a smooth transition from the existing multi-source system to a piped network be guaranteed. In the absence of these necessary conditions, one would expect a simultaneous fall in turnover, and a surge in volume consumed – if service levels are capable of meeting the demand. It is only in their 1990 publication on Nsukka district that the critical importance of volumetric pricing is recognized, following trenchant criticism of the performance of the Anambra State Water Corporation's cost recovery practice.

13.6 CONCLUSIONS

This paper provides a critical review of the early publications of what I call the water demand school with respect to their behavioural studies of the domestic demand for water and waste water services in Ukunda (Kenya), Onitsha and Nsukka district (Nigeria) and Kumasi (Ghana). The school's work has resulted in a permanent transformation of the way the water resource professions approach the domestic sector in the developing countries and has had a major impact on policy formation in institutions such as the World Bank.

The strengths of the work of Whittington et al. and their collaborators are four-fold. They demonstrate how socially complex are the access/distribution networks in the villages and towns of Africa. They give a rich account of volumetric flows, unit prices and turnover along each branch of the network and indicate the dynamism and competitiveness of local water markets. They provide data on the ratio of water service outlays to household income which knock out expectations that these ratios should not exceed an affordability quotient of 5%. They extend their source/distribution network analysis with great success to waste water disposal with respect to volume, unit price and turnover.

The weaknesses of the water demand school are also four-fold. First, to understand the domestic demand for water we need to know how it is used in the household; but these seven studies give little account of how their survey respondents consume the water they access at such great cost relative to their incomes. Secondly, the authors take a naturalistic standpoint in which the research into these communities' behaviour is limited to the actions of silent men, women and children, moving across a landscape like so many ants in a natural-historical study of a savannah colony. Household members are not treated as intelligent, resourceful, purposive and reflexive citizens. Thirdly, the conditional multinomial logit cabbala chosen for the econometric work is hardly transparent, side-steps traditional and justifiable interest in demand functions and their elasticities, and abstracts completely from the uses of water. Fourthly, in their policy work in Ukunda, Onitsha and Kumasi, the water demand school seems to forget that amongst the fundamental attributes of the systems they describe are: costs to the user which are volumetrically determined, and immediate closure of access to water in the case of a household which is unwilling or cannot meet those supply costs.

The principal conclusions of this paper are that behavioural studies into the domestic demand for water and waste water services in the low-income countries should be set within the complex access/distribution networks of a quantitative, whittingtonian type and the price information that these embody; this research should be based on semi-structured interviews, usually with adult females, seeking from these reflexive subjects their account of the origins and character of their service-access routines; the investigation should incorporate the scale and composition of use and reuse and their relation to the quality of water; and the econometric work should act as a servant to this methodological approach.

The material in this chapter originally appeared in: Behavioural studies of the domestic demand for water services in Africa. *Water Policy*, 4(1): 69–81, 2002.

14

Deconstructing households' willingness-to-pay for water in low-income countries

14.1 INTRODUCTION

During the early 1980s growing disquiet began to be expressed over the outcomes of publicly-funded water and sanitation programmes for households living in low-income countries. Many systems had fallen into disrepair and had been abandoned; others were used by only a small part of the original target population; the proportion of costs recovered was low; and expansion to meet the needs of growing populations was too slow (MacRae & Whittington 1988: 247).

As time passed, a new paradigm was broached and then developed by what has been called the water demand school. It suggested that embodying the expressed preferences of individual households is critical to successful project design. It rejected the view that households can afford to pay no more than 3–5% of their income on fresh and waste water services. The new approach highlighted the widespread existence of complex and vigorous water markets. A disciplinary shift took place away from sanitary engineering with its supply-side emphasis to the economic analysis of the demand for water. Finally, the new paradigm suggested that domestic supply programmes could be sustained and replicated only if they responded to the potential *market* for water services.

The mood of the time is well-illustrated in the following quotation (Whittington et al. 1990a: 294):

> If rural water projects are to be both sustainable and replicable, an improved planning methodology is required that includes a procedure for eliciting information

> on the value placed on different levels of service, and tariffs must be designed so that at least operation and maintenance costs (and preferably capital costs) can be recovered. A key concept in such an improved planning methodology is that of 'willingness to pay'. If people are willing to pay for the full costs of a particular service, then it is a clear indication that the service is valued (and therefore will most likely be used and maintained) and that it will be possible to generate the funds required to sustain and even replicate the project.

The purpose of the present paper is to critically review the willingness-to-pay concept's use in the 11 studies listed in Table 14.1 and thereafter to set out the research and policy implications of this critique. As the table shows, the case studies were carried out in the Caribbean, Africa and Asia over the 9-year period from 1986 through 1994 and published in various academic journals in 1988 through 1998. What they have in common, in addition to addressing households' willingness-to-pay for water services in the low income countries, is that Dale Whittington of the University of North Carolina is one of the joint authors in every case. Without a shadow of doubt Whittington is the doyen of this branch of scholarship and so these eleven studies provide a comprehensive account of the new paradigm, making them an appropriate subject for critical review.

14.2 SURVEY METHODS

Before initiating the main argument, it is useful here to sketch the lineaments of the willingness-to-pay surveys carried out in Laurent, Onitsha, the Punjab, Nsukka, Kumasi, Davao, Calamba and Lugazi (see Table 14.1). The pre-requisites of a survey are a budget, the professional staff to plan and carry out the work, identification of the study area, the recruitment and training of local staff as enumerators, and securing whatever institutional support is necessary for the research to go forward. With the team in place, the interview questionnaire is designed and pre-tested, usually

Table 14.1 Eleven studies of the willingness-to-pay for domestic water.

Authors	Case study location	Year of field work	Year of publication
MacRae, Whittington	Laurent, Haiti	1986	1988
Whittington, Briscoe, Mu, Barron	Laurent, Haiti	1986	1990
Whittington, Lauria, Mu	Onitsha, Nigeria	1987	1991
Altaf, Whittington, Jamal, Smith	Punjab, Pakistan	1988	1993
Whittington, Okorafor, Okore, McPhail	Nsukka, Nigeria	1989	1990
Whittington, Smith, Okorafor, Okore, Liu, McPhail	Nsukka, Nigeria	1989	1992
Whittington, Lauria, Wright, Choe, Hughes, Swarna	Kumasi, Ghana	1989	1993
Choe, Whittington, Lauria	Davao, Philippines	1992	1996
Whittington, Choe, Lauria	Calamba, Philippines	1992	1997
Davis, Whittington	Lugazi, Uganda	1994	1998
Whittington, Davis, McClelland	Lugazi, Uganda	1994	1998

Source: see bibliography.

drawing on discussions with local families or community leaders. The sampling technique is explored and selected; this is always *ad hoc* because the statistical universe of households is never known. Inevitably, the sample does not conform to the rigorous demands of statistical theory. However, the view is rightly taken that it is better to have surveyed imperfectly than never to have surveyed at all.

The individual interview may be a once-for-all event; or an initial interview can be followed a day or two later by the collection of answers to specific questions where the household is given time to think more carefully about their response; or the interview is followed up later to give respondents the chance to revise their first reply to key questions. Close supervision of the enumerators is advisable. In Lugazi the survey was followed by a community meeting to discuss the issues raised in the questionnaire (Davis & Whittington 1998). Fieldwork duration varied widely: from as little as two weeks in Onitsha and Lugazi to five months in Kumasi. The number of households interviewed ranged from 170 in Laurent to 1200 in Kumasi. The population size of the survey areas was as little as 1500 persons in Laurent, up to 700,000 in Onitsha where the 'rapid reconnaissance' survey period was 10 days (Whittington et al. 1991: 180–2). In a World Bank review of techniques for systematic client consultation, based on a background paper by Whittington and Davis, 2–4 weeks' fieldwork is recommended (Owen 1994: 16).

The survey questionnaires differed between case studies, of course, but their content can be considered to fall into six categories. *First* is the date of the interview and the address of the household or its location in the study area. *Secondly*, we have the age, gender and household status of the respondent. *Thirdly*, household information would be collected such as the number of persons and their age/gender, their health, education and occupation, and household income, expenditure and assets. *Fourthly*, data are collected on the quality of the dwelling, whether it is owned or rented, and its costs. *Fifthly* comes the sources of the water used in the dwelling, its quality (but not its quantity), conflicts over access, seasonality of supply and demand, the price or charge payable for each separate fresh or waste water source used, household expenditure on water per time period, water storage facilities in the dwelling, current sanitation facilities, the uses to which fresh water is applied, and the degree of satisfaction with the existing service. *Sixthly*, a 'bidding game' (or a referendum technique) would be used to establish the household's willingness-to-pay for alternative water or sanitation projects and this would require some kind of description of the new scenario.

14.3 SIGN AND BEHAVIOUR

The deconstruction of the 'willingness-to-pay' can now get under way. The principal propositions to establish immediately concern critical ontological features of the work reviewed, that is, the 11 studies of Table 14.1. The (im)precise meaning of the willingness-to-pay will be addressed in Section 14.4. But whatever sense is agreed upon, it should be clear that the willingness-to-pay captured by Whittington and his colleagues in their fieldwork is a *sign*, or rather, a collection of signs. These are: the oral reply or replies given by the respondent during the bidding game; the written record of these replies noted down by the enumerator on his answer sheet; the digital information stored on the professional's laptop which becomes manifest on his screen, after transcription from the answer sheet; and the printed information produced in the team's reports and any subsequent publications.

At a quite separate ontological level we have the *behaviour* of the household in accessing water or in using sanitation facilities at some later date in those cases where the project scenario set out in the original interview is implemented. Our specific interest is in the actual payments made by households for those realised projects' fresh and waste water services.

So we have sign and behaviour. I believe that the most interesting and important relationship in this policy field is the degree of consonance which exists between the willingness-to-pay and market behaviour in those cases where scenario projects are in fact implemented. The remainder of this paper explores the conditions which must be met for such sign-behaviour consonance to exist.

14.4 DEMAND THEORY AND SURVEY PRACTICE

The first condition for consonance between sign and behaviour, word matched by deed, is that the willingness-to-pay for a projected fresh or waste water service is both a meaningful concept from a social science perspective *and* that the survey respondent understands that meaning. In economic analysis, whether of the institutional school (within which this paper is located) or the neo-classical school, the demand for a water service can be defined as the relationship, at a given time and within a defined market, between price per unit of the service and the quantity in each time period that consumers would purchase at each price. Demand is, of course, conventionally represented graphically with price on the vertical axis and quantity on the horizontal axis, showing the difference in quantity purchased at each price. Almost invariably higher prices are associated with lower quantities and so the demand function slopes downwards from left to right; it can be represented as a linear, log-linear, exponential or cubic function. The cultural and economic context of demand can be analysed under three rubrics. The first of these is the tastes and habits of consumers, that is, the nature of need for the service or product. The second is the price, quality and availability of services and commodities that consumers consider to be substitutes. These two together should account for the consumers' *willingness*-to-purchase. The third rubric is the incomes, assets and access to credit of consumers, which account for the *ability*-to-purchase (Robinson & Eatwell 1974: 149, Merrett 1997: 53–8).

There are a number of serious problems in the application of demand theory to the survey practice of Whittington and his colleagues. These are: the use of a *maximum* willingness-to-pay price (wtp_{max}), the lack of clarity in the way affordability is dealt with, the treatment of substitutes, and the relevance of sanctions in the case of non-payment of water service bills. I shall deal with each of these in turn.

In all of the studies listed in Table 14.1, it quickly becomes clear that the object of the bidding game section of the questionnaire is *not* to elicit an answer to a question of the sort:

> 'In the case of project X going ahead, would you be willing to pay price Y per drum for the water supplied?'

In fact, what the bidding game elicits is an answer to a question such as:

> 'In the case of project X going ahead, what is the *maximum* price you would be willing to pay per drum for the water supplied?'

So the entire approach seeks data on wtp_{max}.

There is a puzzle here which, to solve, requires us to distinguish between two fundamental ways in which water service payments can be made. The first is by means of a unit price, that is, a payment based on the quantity of water received, or (say) for each visit to a public latrine. The second is by means of a fixed charge per unit time period for access to a service, such as a yard tap connection to a public supply network. In the first case a maximum price simply has no meaning; the household adjusts downwards the quantity purchased the higher the price. In the second case, however, there certainly *will* be a maximum charge above which the user declines the use of the service. This crucial distinction is not discussed by Whittington and his colleagues with reference to wtp_{max}; moreover, Whittington neglects the distinction in his behavioural studies of household demand in African settlements (Merrett 2001). It would also be theoretically meaningful to ask what would be the maximum unit price a household would pay per month *for a specific volume of water received*. But the water demand school never does this.

As a consequence, wtp_{max} is not a meaningful concept from a social science perspective if it is applied to water demand analysis of projects where payment is based on unit prices. Even in the case of projects based on a fixed charge, where wtp_{max} *is* meaningful, the questionnaire design must elicit answers in terms of a fixed charge, not a unit price. Moreover, if the baseline payment mode is of a unit price type, and the scenario payment is a fixed charge, the capacity of the respondent to respond to the wtp_{max} questions in the bidding game will be diminished because of the unfamiliarity of that payment mode for water services (see also Owen 1994: 16).

It is unsurprising then that the World Bank Water Demand Research team states (1993: 49):

> … it was hard to convey the notion of what was meant by the *maximum* an individual would be willing to pay. A respondent in Haiti asked an enumerator, 'What do you mean the maximum I would be willing to pay? You mean when someone has a gun to my head?'.

Elsewhere Whittington himself has made the same point (1998: 22).

It is worth noting with respect to both the baseline and the scenario situations, that the willingness-to-pay literature shows that a key advantage of unit price payments is that they alone give the household a degree of control over its water service expenditure during the course of the year, by means of varying their daily consumption. Such control can be valuable to families whose incomes are low and variable (Whittington et al. 1990b: 1907). It is a curious feature of the water demand school's work that, whilst it has a fascination for the price of water, it says virtually nothing about the quantity purchased by individual households nor how it is used (Whittington & Swarna 1994: 32, 60, Merrett 2001).

14.5 THE AFFORDABILITY QUESTION

The previous section pointed to the conventional distinction between the *willingness*-to-pay for a commodity and the *ability*-to-pay for it. Of course, this is familiar to the water demand school (Whittington 1998: Table 1). However, the distinction is only once made explicit to the households interviewed in the surveys listed in Table 14.1. In practice, the concept of the willingness-to-pay is used by the water demand researchers to mean *both* willingness- *and* ability-to-pay; the separate concepts are rolled up in one.

Unfortunately this opens the real possibility that some respondents will understand willingness-to-pay as inclusive of affordability, and others to take it as exclusive.

So their responses become ambiguous, we do not know quite what they mean. In a 1994 report to the Asian Development Bank, Whittington and Swarna admitted as much in their discussion of field methods that use a question of the type: 'What is the maximum amount of money you would be willing to pay (for a specified good or service)?' They write (Whittington & Swarna 1994: 50):

> If respondents could always provide accurate, reliable answers to such a question, this clearly would be the preferred question format. Unfortunately, for a variety of reasons, this often seems not to be the case. It is often difficult to convey the notion of the 'most' (or the maximum) that one would be (freely) willing to pay, that is, able to pay if willing to do so. Some respondents misinterpret direct, open-ended questions to mean 'What is the most you would *like* to pay?' or 'What is the most you think you *should* pay'. Both of these nuances are clearly not what is meant to be conveyed.

With clarification like this, who needs confusion?

In the field, the research method seems to have been as follows:

- Laurent: an explicit reference to affordability in the bidding game's last question.
- Onitsha, Punjab, Nsukka: no reference to affordability.
- Kumasi, Davao, Calamba, Lugazi: no explicit reference to affordability, but strongly implicit.

14.6 THE TREATMENT OF SUBSTITUTES

In Section 14.4 it was suggested that the first condition for consonance between sign and behaviour requires that the respondent understands the meaning of the wtp_{max} question. It was also pointed out that in demand theory the willingness-to-pay for a service or product is shaped by the price, quality and availability of services and commodities that consumers consider to be substitutes.

In fact, in the low income countries there is often a rich variety of means of access to water services for households. This is illustrated in Table 14.2 for the case of

Table 14.2 Water and money flows in Onitsha – 1987 dry season.

Access branch	Water volume (million gallons per day) (1)	Unit price (naira/gallon) (2)	Turnover (naira)
Household collection from private boreholes	0.30	0.027	8000
Truck sales to households	1.00	0.020	20000
Shop sales to households	1.55	0.050	78000
Doorstep vendor sales to households	0.11	0.127	14000
Shallow well collection by households	0.30	0.000	0
Public piped supply to households	1.50	0.003	5000
Total or weighted average	4.76	0.026	125000

Source: Whittington et al. 1991, Figure 1.
Notes: 1. One imperial gallon = 4.546 litres.
2. In 1987 the rate of exchange of the US$ to the naira was 1:4.3.

Table 14.3 An attribute matrix for household water services.

Source	Price per litre	Payment[1] arrangements	Collection and queuing time per trip	Odour	Turbidity	Other qualities	Pressure	Service times	Service reliability
Household collects rainwater									
Household collects from house well									
Household collects from other well									
Household collects from stand pipe or pump									
Purchase from doorstep vendor									
Purchase from doorstep tanker									
Purchase from shop or kiosk									
Piped supply to house									

[1] Such as a fixed monthly charge or volumetric price collected at the time of sale.

Onitsha. There are six means of access for the city's households. Volume used varies from as little as 0.11 millions of gallons per day (0.5 millions of litres per day) in doorstep vendor sales to 1.55 million (7.05 million litres) in shop sales. The prices households are willing and able to pay vary from as little as zero naira/gallon (n/g) for shallow well collection by households, through 0.003 n/g (0.0007 n/l) for the public piped supply up to 0.127 n/g (0.028 n/l) for doorstep vendor sales, a variation of 1:42 in terms of the positive price range.

This creates some difficulty for the wtp_{max} question. The project to which the question refers can certainly be clearly described using attributes of the type set out in Table 14.3. But the research team is never certain that the project will be *delivered* with the attributes the team describes to the household; and the team is never able to provide an attributes matrix for the substitutes for that project's service.

As a result the respondent household cannot know at the time of the interview the nature of the real choice that will become available to it. It cannot therefore make the kind of commitment necessary for consonance between word and deed. So the 'true preferences' that the water demand school explicitly seeks in response to the wtp_{max} question *do not exist* (Davis & Whittington 1998: 7, Whittington & Choe 1992: 57, Whittington et al. 1992: 212, 1993: 1544).

14.7 SIGN AND SANCTION

There is a special case concerning the meaning of the commitment by a household to pay the wtp_{max}. A household may wish to enjoy a service, be able to afford it *and* be willing to pay for it, yet when push comes to shove, the household does *not* pay for it. This occurs in many, many countries and in my own field experience is widespread in, for example, Armenia, Bangladesh and Peru. The phenomenon is found where there is no sanction for non-payment of the water bill. It invariably takes place only where the service is provided by a public sector institution; private companies that cannot collect on their invoices go out of business.

The optimum conditions for this species of non-consonance to breed and multiply exist: where economic life is hard so that households need to take the greatest care over their domestic expenditure; where there is a widely-held view that certain public services should be free; where persons or parties in political life give their support to non-payment; where the quality of the public service is poor; where the government is so manifestly corrupt that payments for public services are known to line the pockets of the power elite; and where neither the government nor the public water utility is willing to exercise sanctions against non-payment because of the likely political and/or public health consequences.

Two examples are worth citing from the work of the water demand school. In the Nsukka district of Nigeria, the Anambra State Water Corporation had financed borehole drilling but was unable to recover even the operations and maintenance costs of borehole pumping (Whittington et al. 1990b: 1905). In the Punjab non-payment of the flat monthly charge was estimated at 40% but 'Households that do not pay are seldom disconnected from the system' (Altaf et al. 1993: 1905).

In Yerevan, the capital of Armenia with a population greater than one million persons, survey research that I reviewed in 1999 indicates that about 85% of households do not pay their monthly bill to the public sector Yerevan Water and Sewerage Enterprise. Box 14.1 gives a real-life account of the situation of one family I interviewed.

Box 14.1 The price of water in Yerevan.

The official version

Family Alpha lives in Omega district in Yerevan. It is composed of paternal grandmother, father, mother, son and daughter (the respondent). They are owner-occupiers of a small house which is 90 m^2 in size. It has one cold water connection to the Yerevan Water and Sewerage Enterprise supply and the interior points of water use are one flush toilet, one bath, one shower, one kitchen sink, one washing machine, and one wash-hand basin. There is no dish washer and no garden tap. The dwelling receives no hot water supply from any exterior source and is kept warm with an oil heater. Family Alpha also purchases one 0.5-litre bottle of mineral water each day. Their water supply is not metered. They consume 250 litres per head per day, a total for the family of 1250 litres. They pay per month per person 420 drams for their water, 2100 drams per month. The payment is made in cash at the municipal area office. The unit price of water is 55 drams per cubic metre.

The true story

Much of the official version is true ... but there are some twists. There are only three residents because grandmother has moved to the country and the son is living in Moscow. The family has not reported their departure to the Central Police Register because of the legal difficulties this could create for the Alphas. They have estimated for our study that, as best they can guess, the family's real daily consumption is 350 litres, not 1250. Mother is very careful by nature in husbanding the family's resources and sees that everyone else toes the line. So the cost to the family of the fresh and waste water service is, in fact, 196 drams per cubic metre, well over three times the official version. The family is seriously considering going over to a metered supply, especially with rumours that the official price will rise under the new International Water Operator which will begin operations in the year 2000. After moving to the Echmiadzin area, grandmother has decided to stop paying her local bills there: her neighbours say no-one else pays their water bill and there are no sanctions from the water utility. The service is lousy and they are all charged for water they do not get.

14.8 PRIVATE AGENDAS

Sections 14.4–14.7 above dealt with the first condition for consonance between sign and act, that the willingness-to-pay for a projected fresh or waste water service is a meaningful concept from a social science perspective *and* that the survey respondent understands that meaning. The second condition relates to the effects the respondent seeks in giving his or her answer. In a sibling paper to the present one, critically reviewing behavioural studies of the domestic demand for water services in Africa, I have argued that household members interviewed by water resource research teams should be 'regarded as active, intelligent, resourceful, purposive, reflexive subjects, not merely as objects exhibiting quantifiable behavioural patterns' (Merrett 2001). If this approach is accepted, it follows that such subjects will certainly have private agendas of their own when replying to wtp_{max} questions. Indeed, early criticism of what is widely called the contingent valuation approach of willingness-to-pay studies, that it is 'inferior to market data' or 'virtually useless' or that 'responses are self-seeking', reflect a deep-seated concern with private agendas (Whittington et al. 1990a: 294).I suggest that research in this field indicates four categories of respondent in respect of private agendas: the cynic, the strategist, the diplomat and the idealist.

The *cynic* is a person who has little faith in human sincerity and goodness. He (or she) believes the research is a waste of time and the research team a waste of space.

He is likely to refuse the interview, or give an unreasonably low bid price. Cynicism may be specifically directed at government institutions and their performance rather than a sweeping condemnation of all humanity.

The presence of the cynic has been identified by Whittington and his colleagues. For example, the 1986 study of Laurent had a 14% non-response to interview questions on the wtp_{max} for standposts and 25% non-response for private connections (Whittington et al. 1990a: 302). Many of these were probably cynics. In the 1989 Nsukka district study one half of the sample was re-interviewed one day after their first interview in order to see whether they wished to revise their bid. Almost one-fifth did do so, mostly downwards, and these downward revisions lay in the range of a 53–80% cut in the wtp_{max}. The research team took the view that this was due to a lack of trust by respondents in the government water utility to deliver and maintain the project. Indeed, there was an abandoned elevated storage tank and a capped borehole in one of the villages where interviews took place, bearing witness to public sector incompetence (Whittington et al. 1990b: 1906–7, 1992: 219–20). In the 1992 work in Calamba, the researchers suggested that the perceived ineffectiveness of government programmes in the Philippines led respondents to give a *higher* wtp_{max} for projects with a *lower* scope (Whittington et al. 1997: 232).

The *strategist* takes the view that her or his wtp_{max} response carries a commitment to the project and is likely to influence its outcome. She determines that her response should be one most favouring the interests of her family and (perhaps) the local community. Most commonly there is a bias to low prices in these cases, where the aim is to hold down the tariff the water utility will eventually set. However, respondents may also veer to high prices if they believe this is the best way to ensure the project goes ahead.

My reading of the Laurent case study is that downward strategic bias was present: the survey offered a standposts project with no risks for individual households, and they were happy to respond positively to this but offered a low wtp_{max} to keep down the tariff. Similarly in Onitsha, where the scenario was based on volumetric pricing, the wtp_{max} was about 0.2 n/g for the rainy season, less than one-sixth of the price charged by distributing vendors (Whittington et al. 1991: 196). In Nsukka, the time-to-think opportunities brought lower bids; there, strategic behaviour was recognised by the research team as one possible explanation, the others being the search for a just price or the sobering effects of time to consider (Whittington et al. 1992: 220).

The *diplomat* is aware that behind the enumerator lies a group of highly-paid professionals who have flown in perhaps from North America or Europe and who have the support of local bigwigs and the state government. These are people making offers it is unwise to refuse. So the diplomat gives answers which he (or she) believes to be those the enumerator wants and where any hint of what might be a favoured response is rapidly accepted by the interviewee.

The clearest examples of diplomat bias come from surveys in Indonesia (Whittington 1998: 8). Referendum techniques for determining wtp_{max} are especially likely to give evidence of diplomacy. When a referendum elicitation procedure is used, respondents themselves do not need to be asked a question about the maximum they would be willing to pay for a proposed good or service. Instead, split-sample techniques with variation between the samples in the referendum price quoted to the interviewee can be utilised to randomly selected respondents (*ibid.* 22). The statistical distribution of 'yes' answers over the price range can give 'fat tails', that is, an implausibly high number of persons who say that they are willing to pay the higher prices.

The *idealist* is a woman (or a man) who regards the enumerators and the research team behind them as good, serious people; she believes the survey could contribute positively to her family and the local community; accordingly she gives it her best shot, answering all the questions put to her as accurately as they permit. The Laurent survey came across one person of almost Franciscan charity whose wtp_{max} was so high that it exceeded 10% of local annual per capita income, on the grounds not that he would use the new CARE fountains but because 'the community needs it' (MacRae & Whittington 1988: 255).

Contingent valuation studies often have a moralistic tone when referring to the requirement that respondents, having 'true preferences', should reply truthfully and honestly to the questions designed for them by the research group. The working assumption is that those surveyed are idealists but that survey design and socio-econometrics are worthwhile in testing for strategic bias, for example. A more open-minded approach would be to consider respondents as reflexive actors composed of cynics, strategists, diplomats, idealists and, indeed, of persons of mixed affiliation, as was clearly the case in Nsukka. Survey design and analysis, in this case, would include an attempt to identify the proportion of actors who fall into these agenda categories.

14.9 ANCHOR PRICES

In Section 14.3 above it is noted that the willingness-to-pay captured in field surveys is a sign, in contrast to the behaviour constituted by paying for water where a scenario project described in the survey is actually delivered. There it is suggested that consideration of the determinants of consonance between sign and behaviour is the most valuable contribution of methodology, and hermeneutics in particular, in this field of hydroeconomics. Here it is worth stressing that a research team's modelling of its own survey results is *not* a determinant of consonance, for the relationship between word and deed cannot be shaped by such research. For that reason the socio-econometrics of willingness-to-pay studies (the production of signs interpreting signs) is given little attention in this paper; the modelling of wtp_{max} responses deserves a critical review of its own.

The deconstruction in Sections 14.4–14.8 of the willingness-to-pay literature begins by arguing that the willingness-to-pay a maximum price has no locus in the theory of the demand function for water unless it refers to a specific volume received per time period. *Pari passu*, it lacks any clear meaning for a respondent who faces a volumetric pricing tariff for the scenario project. wtp_{max} can have a practical significance only when it applies, for example, to a monthly fixed charge for access to a water supply. Even here the results should be represented as that fixed charge and not as a unit price, for the only sense attributable to the latter in this case is the statistical ratio of the fixed charge to the unknown volume of water used, a ratio the user never computes.

The next attack (in Section 14.5) is that rolling up ability-to-pay and willingness-to-pay together into the wtp_{max} concept, but then rarely clarifying this explicitly, creates for the respondent household an ambiguity at the heart of the survey process. Whittington and his colleagues never say so, but perhaps their unwillingness to make the affordability issue explicit is founded on the fear of thereby encouraging strategic bias.

The third broadside (in Section 14.6) is that the expression of the household's 'true preferences' in a wtp_{max} is not empirically possible since neither the research team nor the household know what substitutes enjoying what attributes may compete with the scenario project over its working life.

The next critique (in Section 14.7) is that, where the appropriate conditions exist, a family may commit itself to a wtp_{max}, desire, be able to afford and enjoy the service ... but not in fact pay for it. The necessary condition for such behaviour is that non-payment attracts no penalty.

The fifth and final blow (in Section 14.8) aimed at the literature under review is that, if we recognize survey households as intelligent, resourceful, reflexive actors, then the response to the questions supposedly eliciting wtp_{max} will be pregnant with the private agendas of the persons interviewed.

Consider a question of the type:

> We have explained our work and described the proposed project; can you now tell us what is the maximum price you would be willing to pay for this service?

If the preceding, five-fold critique of willingness-to-pay studies is well-founded, one would expect that answers to such a question to be the outcome of confusion, uncertainty, disbelief and very private motivations. As a set, the responses would be incoherent, providing no useful wtp_{max} results.

In fact, at least in the published studies, respondents' answers *are* coherent enough to be statistically modelled and tested, for example for the presence of private agendas as well as the variation of wtp_{max} with household income or assets. However, I suggest that the tractability of the data to numerical testing, providing at least some plausible results, springs precisely from the fact that *the type of question posed in the preceding paragraph is never asked*. Instead we have a bidding game where the enumerator presents a plausible wtp_{max} and then gives the household the chance to move above it or below it. Whittington and his colleagues in their publications almost always record the structure of the bidding game (e.g. Whittington et al. 1990a: 302, 1991: 196, 1992: 222–3, 1993: 1557–8).

In the papers listed in Table 14.1 the water demand school repeatedly suggests that a bidding game format is more familiar to respondents because it is like the bargaining process taking place in local markets. This is an absurd claim. Market bargaining concerns a transaction which, once successfully concluded, is immediately implemented; it touches on a commodity lying within the customer's reach; and the potential purchaser never, ever, reveals the maximum price she is willing to pay. The breach of these routines and conventions, with near-fatal results, is the subject of the famous haggling scene published by Chapman et al. (1979).

The use of anchor prices of this type in contingent valuation research, particularly with respect to environmental projects, takes place precisely because, without such a method, the valuation question generates absurd and intractable responses with high variance in the value of the independent variable, wtp_{max}. Elsewhere I have published another critique of the method's application to water resource projects, in that case those seeking to protect the Norfolk Broads from flooding (Merrett 1997: 170–4). There the attack focused primarily on the counterfactual nature of contingent values. It is only a slight exaggeration to say that contingent valuation provides model answers to its own questions. More precisely, in the context of the present paper, such research generates questions eliciting a single, quantitative answer, when the subject demands a complex textual response. In order to successfully elicit a coherent set of signs from the sampled population, the researcher provides the interviewees with anchor prices.

It may be that it was recognition of the leading question character of anchor prices that led some willingness-to-pay studies to try the referendum approach referred to in Section 14.8. The severe drawbacks of the referendum method in low-income countries are exposure to diplomat bias, loss of confidence by respondents and enumerators in the procedure when posing extreme values in the interview, and confusion and even anger in tight-knit communities when different families are assigned widely-varying prices (Whittington 1998: 4–6).

Whittington & Swarna (1994: 8) have written:

> ... the project analyst does not actually have to know why an individual values a project, or even which of the many consequences or effects of the water project that an individual cares about most. It is enough to have a measure of the strength of the individual's preferences for the project; it is assumed that the individual knows his (or her) own interests and is the best judge of what the project is worth to him (or her).

If this is true, why is it necessary to provide 'him (or her)' with an anchor price?

14.10 CONCLUSION

This review of eleven studies in the willingness-to-pay literature is now complete. Next I shall present my own positive views on the demand-side applications of economics to the water resource challenges facing developing country NGOs and governments. In spite of the critical nature of this paper, the policy prescriptions set out below, for the design of fresh and waste water projects aimed at the household sector, do not mark a clean break with 'the new paradigm' described in Section 14.1. Rather, they are a contribution to the evolutionary development of that paradigm.

In the *first* place, the design and development of water and sanitation projects in a local area need to be based upon a good understanding of the existing local market for these services. The water demand school, particularly Dale Whittington, have made a great contribution here to our comprehension of these hydrosocial processes from the point of abstraction to the moment of water purchase by the household. This now needs to be complemented by an attempt to grasp how much water is used by the family, by whom and to what purpose (Merrett 2001). Without this baseline behavioural groundwork, projects in the scenario year are exposed to a greater risk of failure.

Secondly, our understanding of future household behaviour requires semi-structured interviews primarily with families' female adults. It is women who play the primary role in the collection and purchase of water, as well as its internal use in cooking, washing and cleaning. For this reason alone, the professional team should include a female sociologist or socio-economist. In large surveys the bulk of the data collection will be done by trained enumerators (ideally women). But the professionals in the research team should play the central part in at least a sub-sample of the interviews. This also calls for national professionals to form part of the research team. Surveys where enumeration is the sole responsibility of local secondary-school graduates, but where data analysis and report writing is the preserve only of international staff, should be strictly avoided. Furthermore, respondents should be recognised not as objects sourcing numerical signs but as intelligent, purposive, reflexive subjects, rich in their textual accounts of their neighbourhood and their region. One should also appreciate that respondents are made up of persons with a variety of private agendas: cynics, strategists, diplomats and idealists.

Thirdly, the scenario project option(s) can be developed prior to the survey or, with the sampled population, by the research team itself. In either case the expressed preferences of the targetted groups are vital, as the 'new paradigm' insists so eloquently. The project option should be costed in terms of capital investment as well as ongoing outlays. The use of wtp_{max} in the survey questionnaire and the employment of anchor prices would cease, to be replaced by a question of the type:

> We have explained our work and described the proposed project, including the means that households would pay for it. To cover its full costs the project would require a price of *x* naira/gallon. Can you now tell us whether your family would be willing to pay and able to afford *x* naira/gallon?

Where appropriate, 'a monthly fixed charge of *y* naira per family' would replace 'a price of *x* naira/gallon'. If the project would be financed and approved only on the basis of a full-cost tariff, the survey respondent would be informed of this. This approach would face the same difficulties as the current paradigm in characterizing scenario substitutes (*vide* Section 14.6). The final report would stress that the project is likely to succeed financially only if there are penalties for the non-payment of water bills. Points 1–3 suggest the professional team for this twilight zone would be composed of a statistician/data analyst, a female socio-economist and a public health engineer.

The material in this chapter originally appeared in: Deconstructing households' willingness-to pay for water in low-income countries. *Water Policy*, 4(2): 157–172, 2002.

15

Industrial effluent policy: economic instruments and environmental regulation

Falstaff: *Sirrah, you giant, what says the doctor to my water?*
Page: *He said, sir, the water itself was a good healthy water; but, for the party that owed it, he might have more diseases than he knew for.*

William Shakespeare, *Henry IV*, Part Two, Ii

15.1 INTRODUCTION

The uses of outstream water in human society can be classified into three grand orders: by households for domestic purposes; by farmers for the irrigation of land; and by 'industry' in the widest sense of the term. The industrial uses of water include heating and cooling processes, steam production, washing and cleaning, factory-scale cooking, incorporation in the product itself as in beer manufacture, the drinking water of animals in factory farming, the life-medium of fish as in trout farming, hot water disinfection, dyeing, transporting inputs, products and waste within industrial premises, and fire-fighting.

These industrial uses in manufacturing, livestock factories, hospitals, commerce, offices, mining, petroleum refining, power generation, railway companies etc. inevitably generate industrial effluent, that is, aqueous waste. In the 250 years since the Industrial Revolution, such waste has been disposed of predominantly by dumping it untreated into rivers, lakes, the sea and even underground. The environmental impact of the use in this way of the natural world as a sink has been to destroy animal and plant species, degrade habitats and spread illness and death within human populations.

In the quest for a sustainable society, it has been proposed that water resource management comprises six fields of action, one of which is the purification of water prior to and after its use (Merrett 1997: 148–52). An important economic instrument of relevance to such action is disposal charges, also known as discharge fees, for industrial effluent. Bhatia et al. (1994: 3) suggest that the effective use of such charges is rare in developing nations. But they are certainly widely employed in high-income countries. This paper briefly considers the network of institutional, economic and regulatory relationships encompassing effluent generation and disposal, then turns to the objectives and design of discharge fees using examples drawn from the European Union, and finally compares and contrasts the power, precision, and interrelation of discharge fees and environmental licensing.

15.2 THE GENERATION AND REGULATION OF INDUSTRIAL EFFLUENT

The generation and disposal of aqueous waste is illustrated in Figure 15.1. Abstracted water is supplied to industry for the multifarious uses already described and this leads to the production of raw effluent. In some cases this is collected and discharged directly to rivers, lakes, estuaries and coastal waters, or to underground locations. In other cases the waste water is discharged to sewer for treatment in a sewage treatment works. The third possibility is that the effluent receives pre-treatment on the premises and then is disposed of to water course or sewer or is cycled back for internal reuse within the institution whence it came.

With respect to effluent discharged to sewers and collected for transfer to a sewage treatment works, this works has two outputs: sewage sludge and treated waste water. The sludge may be used in agriculture or incinerated or sent to landfill sites or dumped

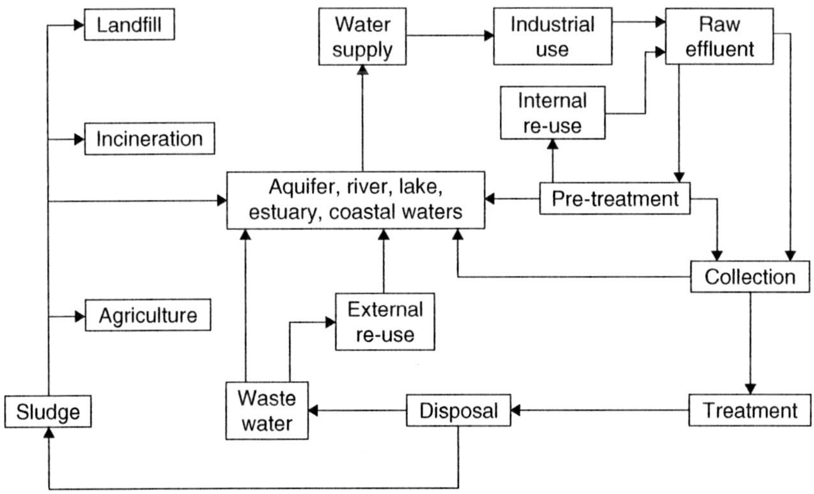

Figure 15.1 The generation and disposal of aqueous waste.

into water courses. Waste water is disposed of to surface waters and the sea. Alternatively it is reused externally in other industrial organizations or as irrigation water.

In the European Union, North America and other regions of the globe, environmental regulators control the disposal of raw effluent, sewage sludge and treated waste water. As already suggested, the polluting content of these solid and liquid wastes has the potential to damage the natural environment and bring illness and disease to urban and rural populations. For example, this is true of complex effluents containing phenols, ammonia, chlorine, heavy metals and organo-phosphates. The government's regulatory agency imposes a ban on some toxins. In other cases it limits their maximum level by means of consent conditions.

Specifically with respect to the European Union, the 1991 Urban Waste Water Treatment Directive lays down uniform emission standards for sewage treatment for all sewage treatment works serving populations of two thousand or more. The Directive also covers industry's direct disposal of effluent. In the UK, the most dangerous substances which can be found in industrial effluent and domestic sewage are on the Environment Agency's Red List.

Environmental regulation raises the cost of pre-treatment by industry and the cost of waste water treatment by water service companies. Industrial companies may therefore oppose the introduction of pollution legislation or may attempt to evade it when it is in place. As a result, the regulator requires systematic monitoring of the chemical composition of discharges to water courses and sewers.

The prohibitions and consent conditions imposed on water company discharges in turn may lead the company to set quality standards on industrial discharges to its sewers. This is because specific industrial effluents can raise sewage treatment works costs substantially, within a defined regulatory framework, or disable some treatment processes. The water service company needs to shield its assets from damage, protect its operatives, keep its production costs as low as possible and ensure its treated waste water discharge and sludge output meets the regulator's standards. As John Hills points out in his valuable book, *Cutting Water and Effluent Costs*, a water service company ban on complex effluents in turn places pressure on industry for strengthened on-site pre-treatment facilities (Hills 1995).

The interdependence of industry, water utility and environmental regulator is readily illustrated. The bulk of sewage sludge in the UK is used on farmland, but heavy metals in the sludge can make it unacceptable for agricultural purposes. Conventional treatment of effluent does not remove these pollutants. As a result, the Environment Agency and the water utilities impose strict limits on the heavy metal content of discharges to water courses and sewers, for example by the electroplating and photographic film processing industries (Hills 1995: 26–7).

So we can see that the introduction of vigorous environmental regulation of industrial effluent disposal to fresh and salt water sinks places new technological demands both on water utility treatment of aqueous waste as well as on the pre-treatment of its effluent by industry itself.

15.3 THE OBJECTIVES OF DISPOSAL CHARGES

A discharge fee on industrial effluent can be defined as a payment levied on an industrial organization by a government regulator or by a water utility in respect of the volume of effluent and/or the pollutant mass it discharges to sewer, aquifer, river, lake, estuary or coastal waters.

To understand the objectives of the institutions which set and collect disposal charges, a vital analytic distinction is necessary based on the flow chart of Figure 15.1. Two types of organization impose discharge fees:

1. Water service companies levy charges on industrial organizations (as well as domestic consumers) which discharge their effluent to the companies' sewers; and
2. the environmental regulator levies charges on industrial organizations, *including water service companies*, which discharge their effluent to water courses.

In the first case above, the utility provides a waste water service which embraces the collection of effluent through the sewerage network, its treatment, and the disposal of waste water and sewage sludge. The objective of the utility in setting the disposal charge is to cover the prime and overhead costs of providing this sanitation service (Merrett 1997: Figure 3.2). Disposal fees of this type are universally referred to as sewage charges. In 1999 a typical charge in the UK is of the order of US $0.80 per cubic metre of waste water.

In the second case, that of the environmental regulator, the objectives are more complex. The control of industrial effluent discharges to water courses imposes a variety of financial costs on the regulatory authority, such as for research, policy-making, the preparation of discharge licences, and monitoring adherence to the standards set. Hötte et al. (1995: 220) indicate that the original objective of the levy on discharges to surface waters in the Netherlands was 'fund-raising for water quality management by government authorities'. We can call this the hypothecation objective. Such a goal is probably the origin of discharge fees in every regional or catchment authority in the world, where regulatory charges are collected. The parallel here with fee payments for abstraction licences is evident (Merrett 1999a).

But another goal for discharge fees seems to have developed in an evolutionary manner from the hypothecation charge. At the most general level, the objective of *regulatory standards* on industrial effluents has been to reduce industrial pollution of the natural environment. This has benefits not only of an environmental nature but also can increase gross domestic product: by reducing the costs of instream water users such as in fishing, leisure pursuits and other ecosystem activities; and by cutting the costs of downstream water abstractors. In applying hypothecation fees, it became evident to regulatory authorities that industry was cutting back on the effluent mass it discharged in order to reduce its exposure to these now internalized costs. Once again, the case of the Netherlands is instructive here. So the charge raised to finance the command and control system was now acting independently as an economic instrument in achieving the prime goal of regulation itself. As a consequence, with the passage of time, the second objective of discharge fees, pollution reduction, became established.

15.4 THE DEMAND FOR WASTE WATER SERVICES

Waste water services are a public good, that is, where the benefit derived from the good by one consumer does not diminish the benefit derived by consumers in general. The standard textbook example is a country's armed forces (Sandmo 1987). In a *laissez-faire* economy, it is impossible to price or to sell services for the treatment and subsequent disposal of industrial effluent, for no purchaser of such services privately appropriates their benefits.

Where the state intervenes to defend the public realm against industrial pollution, a space is opened up by environmental legislation within which sewerage utilities (private or municipal) can operate. As we have seen, sewage charges payable to the utility and discharge fees levied by the regulator for the use of water courses as a sink, create a response from industry.

Organizations confronted by disposal charges on the volume of their effluent and/or on the pollutant mass will investigate ways of cutting the charge payable, provided that the fees are not negligibly low. This requires operational research combining managerial, engineering and economic skills. The relation between charge and industrial action can be measured by the elasticity of response of the pollutant mass/discharge fee function (see below).

One method is to reduce the total volume of abstracted water consumed or the total mass of a specific chemical used for the production process. For example, a site survey can be used to calculate the organization's own 'regional' water balance statement, to revise water need calculations and to check the functioning of the water distribution system so as to cut internal water supply volumes and reduce on-site leakage (Merrett 1999b). Another example, from the leather industry, is advances in the technology of the tanning process which permitted a fall in chromium use because of an increased percentage uptake in the hides.

A second method for industry to reduce its disposal charges is pre-treatment by the institution of its own effluent so as to diminish the pollutant mass discharged to sewer or to water course. Pre-treatment includes air flotation for fats, special separators for oil, and bacterial processes for organic effluents. Pre-treatment is likely to be introduced only if the financial rate of return is attractive when its costs are assessed against the reduced bills for disposal. A potentially beneficial spin-off from pre-treatment is by-product manufacture for sale to niche markets, such as the production of animal feed in effluent treatment by whisky distillers.

The third method for cutting discharge fees is for industry to reuse its water and/or its process chemicals. Often this demands pre-treatment as a first step. The impact on water volumes of internal reuse is both to cut its water intake requirements, with an associated fall in water supply costs, as well as to reduce its effluent volume, thereby diminishing its disposal charges. These opportunities frequently occur in process washing where, as Hills points out (1995: 85): 'the cleanest water washes the nearly clean product and the reused water washes the dirty incoming product'. With respect to a process chemicals example, equipment exists to handle the rinse waters from chrome and nickel electroplating drag-out tanks, enabling the plating chemicals to be returned to the plating bath.

The effect of disposal charges on industrial firms' behaviour is illustrated in Figure 15.2. The vertical axis measures the pollutant mass discharged per unit time period, that is, the variable M of equation (1) below. The horizontal axis measures the independent variable P, the price charged as a disposal fee per unit of pollutant mass. The shape of the function is that of a quadratic equation:

$$M = aP^2 + bP + c \qquad (1)$$

where a, b and c are the equation's parameters and a is negative.

Figure 15.2 embodies the hypothesis that at low prices the mass discharged is virtually unaffected by price differences. Here managers take no interest in the invoices

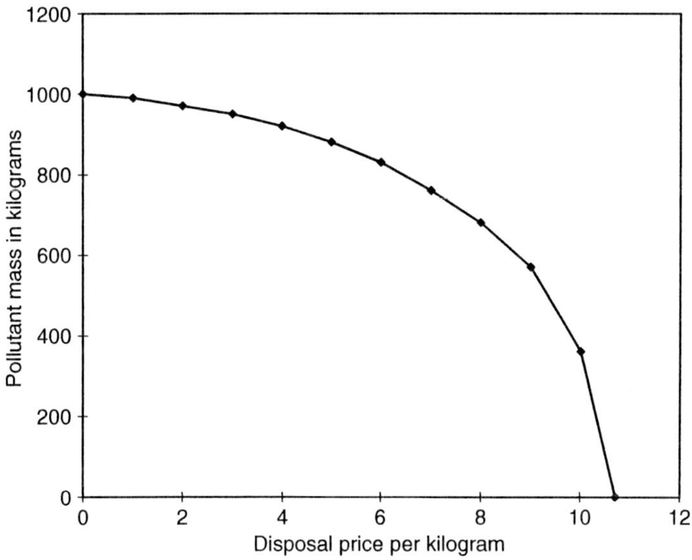

Figure 15.2 The effective demand for pollutant disposal services.

they receive from the environmental regulator or the water utility. In a higher price range, the discharge fee really bites and the slope of the curve is steeper. At very high prices, pollutant discharge ceases as the industrial institution is forced for financial reasons either to change its technology or discontinue production. The price at which M falls to zero has the same outcome as that where a regulatory agency places a complete ban on the pollutant's discharge.

15.5 THE MEASUREMENT OF POLLUTION

Before an understanding of the design of discharge fees is possible, it is necessary to examine how pollutant flows are calculated. One can only price what one can measure. Clearly, the starting point is the flow volume of industrial effluent, for example in cubic metres discharged per day. A second measure is the concentration of pollutants in the effluent. This can be measured as the mass of a *specific* pollutant per unit volume, for example measured in grams per cubic metre, equal to milligrams per litre. This gives:

$$M = V \times C \tag{2}$$

where M is here measured in grams/day, V is effluent volume and C is pollution concentration.

However, pollutant-specific fees entail high transaction costs because of the technical difficulties of measuring the pollutant concentration and monitoring how it changes over time (Green 1990: 9). Effluent is a complex waste. In the steel city of Jamshedpur in India, for example, Bhatia et al. (1994: 10–11) showed that untreated industrial discharges into the Subernarekha River contained phenols, oil, grease, ammoniacal nitrogen, cyanide, chromic and other acids, and hexavalent chromium. The discharge volume is so high that in the low flow month of April the ratio of the effluent volume to the river flow is 1:3.75.

These inquinating cocktails need grouping categories to facilitate the measurement and pricing routines, such as biological oxygen demand (BOD), chemical oxygen demand (COD) and suspended solids, all quantifiable in grams/cubic metre. BOD and COD are directly related to dissolved oxygen, the principal measure of water quality in terms of its ability to sustain plants, fish and other biota.

In France, Tuddenham (1995: 201–2) reports that pollution charges are based on the quantity of pollution generated on an ordinary day in the month of maximum discharge. There are eight grouping categories: suspended solids, oxidizable substances, soluble salts, nitrogen, phosphorus, organohalogenated compounds, metals and metalloids, and toxins. The first seven are measured in grams and the eighth in toxic units. In the Netherlands, the principal measures are COD and the heavy metal pollutants cadmium, mercury, arsenic, copper, nickel, zinc and lead (Hötte et al. 1995: 221–2).

15.6 THE DESIGN OF DISPOSAL CHARGES: THE UTILITIES

Discharge fees should be designed to achieve the objectives pursued in setting them. For that reason, one must deal separately with the water service company and the environmental regulator.

The water utility seeks to cover the costs of providing its sewerage service to industry, such costs defined in a broad sense to include the cost of capital. Its prime costs per unit output comprise wages, salaries, power, materials, spare parts and other consumables. The gross margin per unit output meets overhead costs and net profit. Overheads are made up by the rent of land and buildings, leasing costs, payments to directors, interest on loans and amortization. Net profit is split between taxes, dividends and retained profits.

Pricing strategy is likely to be based on average cost, as Lee (1994) proposes. I have suggested elsewhere that the most appropriate average cost model for the waste water industry is mark-up pricing. In this case average prime cost per cubic metre of industrial effluent at normal capacity is calculated and a mark-up added (Merrett 1997: 59–61). The mark-up may be in cents per cubic metre or it may be a percentage addition to prime cost. Where prices are controlled by a regulator, these too will be average cost based.

There is no space here for prices set at short-run marginal cost nor should there be. Similarly, there is no role for the average total cost of new investment in network and headwork infrastructures, otherwise known as long-run marginal cost. Short-run marginal cost is negligibly low and pricing on this basis would ruin the utility. The average total cost of additional capacity as a price determinant would generate huge losses (or surpluses) wherever average total cost was substantially lower (or higher) than the average total cost of existing capacity. Marginal cost pricing, if combined with a standing charge, discriminates against small-scale users of the sewerage system. This approach to cost recovery pricing in the collection and treatment of waste water, embracing both prime and overhead costs, is embodied in the European Union's Water Framework Directive, which also requires the design of the most cost effective measures to achieve local reductions of substances prioritized on the basis of risk (Bjerregaard 1998: 78–80).

The starting-point for industrial disposal charge setting will be the average cost per cubic metre of collection, treatment and disposal of *all* sewage. Where the concentration of pollutants from specific organizations is calculated to be higher than the average and where this difference gives rise to higher treatment or disposal costs, a higher price should be set, such that the resulting price difference matches the higher cost, on the basis of cost engineers' calculations.

An example of such customary practice is the modern Mogden formula (Mogden is the largest sewage treatment works in London):

$$K = R + Q + B(O_t/O_s) + S(S_t/S_s) \tag{3}$$

where K is the charge per cubic metre; R are the costs for conveyance through sewers; Q is the volumetric treatment cost covering screening, primary settlement, tertiary treatment, and outfalls of treated sewage; B is the cost of biological treatment plus a proportion of secondary sludge treatment and sludge disposal; S is the cost of treatment and disposal of primary sludges; O_t/O_s is the COD in milligrams per litre after 1 hour's settlement at pH 7 of the company's industrial effluent divided by the regional average of the same measure; and S_t/S_s is the suspended solids in milligrams per litre after 1 hour's of settlement at pH 7, again of the company's industrial effluent divided by the regional average of the same measure. Here, then, the multipliers are based on the concentration of chemical oxygen demand and suspended solids. In the UK, Stewart has successfully developed a logarithmic model of the functional expenditure of large sewage treatment works that shows clear economies of scale and uses separate terms for maximum BOD and ammonia consents (Ofwat 1994).

15.7 THE DESIGN OF DISPOSAL CHARGES: THE ENVIRONMENTAL REGULATOR

The regulator's two objectives in charge-setting, discussed in Section 15.3 above, are hypothecatory and pollution-reducing. These will be reviewed in turn. But first it is necessary to make a simple point, often neglected, with great bearing on the regulator's charges. At the simplest level, industrial effluent is composed of two parts: water and pollutants. The ratio of the first to the second is large. For example, Tuddenham produces data that show that in France the mass of pollutants produced per inhabitant per day is 166 grams (1995: 203). The volume of water used per inhabitant per day is of the order of 150 litres. Therefore the ratio of the total mass of pollutant to the mass of water is about 1:900. The discharge of water to river, lake and estuary imposes no costs on the environment. Indeed, it is a welcome addition within the hydrosocial cycle to the catchment's effective rainfall. In contrast, pollutants do impose such costs. Thus the regulator's golden rule should be to levy disposal fees only on the pollutant mass, never on the water volume.

The hypothecatory objective requires that total receipts from the discharge fees levied on industry should equal the total costs of the regulation of industrial effluent. In practice, these two variables are interdependent. The first determines the second which determines the first. To understand the contingencies of their relation requires an economic history of the catchment and an account of its environmental politics.

The French basin agencies provide a clear case of the hypothecatory principle (Tuddenham 1995).

The pollution-reducing objective implies that the fee per unit measure of pollution (see Section 15.5) should vary with the environmental cost it is assessed to impose. In some cases it may be possible to measure the financial cost suffered by in-stream uses such as navigation, recreation and fishing, or by organizations abstracting water for subsequent use. However, I know of no case where such measured financial cost has in fact shaped the discharge tariff. In any case, the principal harm done by pollution is perceived to be environmental, rather than GDP-reducing. In my view such impacts cannot usefully be measured in dollar terms (Merrett 1997: 168–70). In such cases the design of the components of the tariff is best carried out on the basis of environmental impact assessments. Such impacts vary with pollutant type, the ecology of the water course at the point of disposal and downstream thereof, the quality and volume of the river flow at the point of disposal, the season, and the concentration of pollutants within the effluent discharged.

Finally, in this section, it is appropriate to refer briefly to a pricing mechanism for the release of pollutants to water course sinks which environmental regulators may wish to consider. This is the introduction of tradeable disposal rights, a parallel concept to tradeable abstraction rights. Such an arrangement would see the regulator auction permits to pollute to the highest bidder. Such rights would be recognized, for example, for 5 or 10 years and would state the maximum mass of pollutants permitted to be discharged. They would be tradeable between organizations.

Unfortunately, the polluting institutions of industry each has a pollutant mass they wish to discharge annually which differs in total mass, pollutant composition, timing of discharge and location of disposal. Thus it would be impossible for a regulator to design a permit with a known environmental impact which would be of value as a commodity, for it could not be of commercial interest to more than one institution. As Bhatia et al. write: 'effluent permit trading schemes have been implemented ... only rarely for water pollution' (1994: 7).

15.8 CONCLUSIONS

On the basis of the preceding discussion of the disposal of industrial effluent to water courses, there seem to be eleven conclusions an environmental agency might draw on the inter-relation of environmental regulation and economic instruments in the quest for a sustainable society.

1. In terms only of effluent discharged, disposal fees in the course of time can have the same broad outcome as regulatory limits or bans.
2. Both regulation and discharge fees stimulate pre-treatment and reuse within industry, the regulation of industrial discharges to sewer by the water service companies, and specific forms of sewage treatment by those same utilities.
3. Both regulation and disposal charges engender substantial transaction costs for the environmental agency in monitoring industry and waste water utility behaviour. This is because waste water treatment and the disposal of its products are costly public goods. Grouping categories for the measurement of industrial pollution, such as biological oxygen demand, help reduce these transaction costs for both approaches.

4. For both command and control as well as economic instruments, the golden rule must be to target pollutant mass, or its derivatives such as chemical oxygen demand, never the volume of industrial effluent, in order to reduce pollution without diminishing the recycling of water.
5. The design of regulatory proscriptions and consents, as well as of discharge fees, should be based on environmental impact assessment, not on vain attempts to give a meaning to the dollar valuation of environmental costs.
6. Disposal charges generate a hypothecated income for the environmental agency. Regulation does not.
7. Regulation has precision, speed, directness and forcefulness unmatched by discharge fees.
8. The prime and overhead costs of the collection, treatment and disposal of industrial effluents can be prodigious and should be carefully reviewed before regulatory bans and consents are set (Briscoe 1995).
9. Disposal charges permit a flexibility of response by industry unmatched by regulation. This flexibility can be manipulated by the agency so that the industrial costs of conforming to required standards are minimized.
10. The objective of regulation has always been to reduce the industrial pollution of the environment, whereas the objective of discharge fees has been hypothecatory. However, in the course of time, discharge fees have evolved so as also to serve a pollution-reducing objective.
11. Regulation works best where draconian and absolute action is required. Economic instruments work best where partial restrictions shared amongst a group of stakeholders are appropriate. and where limits on pollution behaviour are more costly to society if undertaken by some organizations rather than by others. Under these assumptions, economic instruments can allocate the permission partially to pollute in the most efficient manner.

In conclusion, my judgment is that both approaches are valuable and should complement each other, provided that they are introduced in a society where the design of rules and of prices are appropriate to their objectives, where it is possible to implement regulation effectively, and where disposal charges can be correctly assessed and are actually paid.

The material in this chapter originally appeared in: Industrial effluent policy: economic instruments and environmental regulation. *Water Policy*, 2, 201–211, 2000.

Bibliography

Abernethy, C.L. (2003). Financing river basin management. In: Svendsen, M., Merrey, D.J. & Shah, T. (2003). *Irrigation and River Basin Management: Options for Governance and Institutions*. Wallingford: CABI.

Allan, T. (1995). 'The political economy of water: reasons for optimism but long term caution'. In: Allan, J.A. & Court, J.H.O. (eds) *Water in the Jordan Catchment Countries: a Critical Evaluation of the Role of Water and Environment in Evolving Relations in the Region*. London United Kingdom: School of Oriental and African Studies.

Allan, T. (2001). *The Middle East Water Question: Hydropolitics and the Global Economy*. London: Tauris.

Allan, T. (2003). 'Virtual water – the water, food, and trade nexus useful concept or misleading metaphor?'. *Water International*, 28(1): 106–113.

Allen, R.G., Willardson, L.S. & Fredricksen, H.D. (1997). Water use definitions and their use for assessing the impacts of water conservation. In: de Jager, J.M., Vermes, L. & Ragab, R. (eds) *Sustainable Irrigation in Areas of Water Scarcity and Drought*. Oxford: ICID.

Altaf, M.A., Whittington, D., Jamal, H. & Smith, V.K. (1993). Rethinking rural water supply policy in the Punjab, Pakistan. *Water Resources Research*, 29(7): 1943–1954.

Anderson, P. (1997). Demands and resources for public water supplies in Scotland from 1991 to 2016, in *Water and Environmental Management*, Vol. 11, No. 3, pp. 164–169.

Anglian Water Services Limited (2002). *Annual Report and Accounts 2002*. Huntingdon, AWSL.

Anglian Water Services Limited (2003). *Anglian Water Services Drought Plan March 2003*. Huntingdon, AWSL.

Berkoff, D.J.W. (1990). *Irrigation Management on the Indo-Gangetic Plain*. Washington DC: World Bank.

Berkoff, J. (2003). *Prospects for Irrigated Agriculture: Has the International Consensus Got It Right?* http://www.brad.ac.uk/acad/bcid/GTP/altwater.html.

Bhatia, R., Rogers, P., Briscoe, J., Sinha, B. & Cestti, R. (1994). *Water Conservation and Pollution Control in Indian Industries: How to Use Water Tariffs, Pollution Charges, and Fiscal Incentives*, Washington: World Bank.

Bjerregaard, R. (1998). Getting Europe's waters cleaner: getting the citizens involved, *Water Policy*, 1(1): 73–80.
Blackie, J.R., Houghton-Carr, H.A., McCartney, M.P., Moores, J.P. & Davies, H.N. (1996). *The Trinity Catchment Study: Final Report*, Wallingford: Institute of Hydrology.
Blomqvist, A. (1996). *Food and Fashion: Water Management and Collective Action among Irrigation Farmers and Textile Industrialists in South India*. Linköping: University of Linköping.
Briscoe, J. (1995). *The German Water and Sewerage Sector: How Well it Works and What This Means for Developing Countries*. Washington: World Bank.
Bruns, B.R. & Meinzen-Dick, R.S. (2000). *Negotiating Water Rights*. London: IT Publications.
Bruns, B.R. & Meinzen-Dick, R.S. (2001). Water right and legal pluralism: four contexts for negotiation. *Natural Resources Forum*, 25: 1–10.
Cairncross, S. & Kinnear, J. (1991). Water vending in urban Sudan. *Water Resources Development*, 7: 267–273.
Cairncross, S. & Kinnear, J. (1992). Elasticity of demand for water in Khartoum, Sudan. *Social Science and Medicine*, 34: 183–189.
Chapman, G., Cleese, J., Gilliam, T., Idle, E., Jones, T. & Palin, M. (1979). *Monty Python's The Life of Brian (of Nazareth)*. London: Eyre Methuen.
Chilton, P.J. & Bird, M.J. (1994). *Nitrate in Jersey's Groundwater*. Keyworth: British Geological Survey.
Choe, K., Whittington, D. & Lauria, D. (1996). The economic benefits of surface water quality improvements in developing countries: a case study of Davao, Philippines. *Land Economics*, 72(4): 519–537.
van Damme, H.M.G. & White, A. (1984). Technology choices for the decade. In: Bourne, P.G. (ed.) *Water and Sanitation: Economic and Sociological Perspectives*. San Diego, Academic.
Davis, J. & Whittington, D. (1998). 'Participatory' research for development projects: a comparison of the community meeting and household survey techniques. *Economic Development and Cultural Change*, x (y): 73–94.
Department of Agriculture and Fisheries (2002). *Statistics 2001*, Trinity: DAF.
Dinar, A. (ed.) (2000). *The Political Economy of Water Pricing Reforms*. New York: Oxford University Press.
Dow, S.C. (2002). *Economic Methodology: an Inquiry*. Oxford: Oxford University Press.
Dubourg, W.R. (1993). *Sustainable management of the water cycle in the United Kingdom* (Norwich, Centre for Social and Economic Research on the Global Environment).
Dubourg, W.R. (1995). *Pricing for sustainable water abstraction in England and Wales: a comparison of theory and practice* (Norwich, Centre for Social and Economic Research on the Global Environment).
Dyson, J.R. (1994). *Accounting for non-accounting students*. London, Pitman, 3rd edn.
Earl, P.E. (1995). *Microeconomics for Business and Marketing*. Cheltenham: Edward Elgar.
Environment Agency (2002a). Demand Management Bulletin, 54 (August): 3.
Environment Agency. (2002b). *Drought Plan Guideline*. Bristol: Environment Agency.
Environment Agency Anglian Region (2001). *Water Resources for the Future: a Strategy for Anglian Region*. Peterborough: Environment Agency.
Environment Agency Anglian Region (2003). *2003 Regional Drought Management Plan – Final Plan*. Peterborough: Environment Agency.
European Commission (1991). Urban Waste Water Treatment Directive 91/271/EEC, Brussels: EEC.
Fass, S. (1988). *Political Economy in Haiti: the Drama of Survival.* (New Brunswick NJ: Transaction Books).

Foster, S., Adams, B., Morales, M. & Tenjo, S. (1993). *Groundwater protection strategies: a guide towards implementation* (Lima, CEPIS).

Fowler, S. (1995). *Water wise: the RSPB's proposals for using water wisely* (Sandy, Royal Society for the Protection of Birds).

Frederiksen, H.D. (1996). Water crisis in developing world: misconceptions about solutions. *Journal of Water Resources Planning and Management*, 122(2): 79–87.

Frigot, D. (2001). Jersey's Rural Heritage: a Farming Way of Life, Jersey: Royal Jersey Agricultural & Horticultural Society.

Gale, R., Barg, S. & Gillies, A. (1995). *Green Budget Reform: an International Casebook of Leading Practices*. London: Earthscan.

Gass, G.M., Robins, N.S. & MacDonald, A.M. (1996). Groundwater Resources Degradation in Jersey: Socio-Economic Impacts and their Mitigation, Keyworth: British Geological Survey.

Gazmuri Schleyer, R. & Rosegrant, M.W. (1996). Chilean water policy: the role of water rights, institutions and markets. *Water Resources Development*, 12(1): 33–48.

Giddens, A. (1984). *The Constitution of Society: Outline of the Theory of Structuration*. Cambridge: Polity Press.

Gleick, P.H., Loh, P., Gomez, S.V. & Morrison, J. (1995). *California water 2020: a sustainable vision* (Oakland, Pacific Institute for Studies in Development, Environment and Security).

Green, C.H. (1990). *Economics in a Time of Failure*. London, Middlesex Polytechnic, Flood Hazard Research Centre Publication No. 180.

Haddad, B.M. (2000). *Rivers of Gold: Designing Markets to Allocate Water in California*. Washington DC: Island Press.

Hearne, R.R. & Easter, K.W. (1995). *Water Allocation and Water Markets: an Analysis of Gains-from-Trade in Chile*. Washington DC: World Bank.

Hills, J.S. (1995). *Cutting Water and Effluent Costs*, Rugby: Institution of Chemical Engineers.

Hodge, C.A.H. & Seale, R.S. (1966). *The Soils of the District around Cambridge*. Harpenden: Agricultural Research Council.

Hodgson, G.M. (1994). Cultural and institutional influences on cognition. In: Hodgson, G.M., Samuels, W.J. & Tool, M.R. (eds) *Institutional and Evolutionary Economics A-K*. Cheltenham: Edward Elgar.

Hodgson, G., Samuels, W. & Tool, M. (eds) (1994). *The Elgar Companion to Institutional and Evolutionary Economics*, Cheltenham: Edward Elgar.

Hoekstra, A.Y. & Hung, P.Q. (2002). 'Virtual water trade: a quantification of virtual water flows between nations in relation to international crop trade', *Value of Water Research Report Series* No 11. Delft: The Netherlands: IHE.

van Hofwegen, P. & Svendsen, M. (2000). *A Vision of Water for Food and Rural Development*. The Hague: Netherlands Directorate General for Development Cooperation.

House of Lords (2003). *Water Bill [HL]*. London: Stationery Office.

Howsam, P. & Carter, R. (eds) (1996). *Water policy: allocation and management in practice* (London, Spon).

Hötte, M.H.H., van der Vlies, J. & Hafkamp, W.A. (1995). Levy on surface water in the Netherlands, in Gale et al. *op.cit*. pp. 220–230.

Jersey New Waterworks Company Limited (2002). Annual Report and Accounts for the Year Ended 31st December 2001, St Helier: The Jersey New Waterworks Company Limited.

Kaminsky, G., Lizondo, S. & Reinhart, C. (1998). Leading indicators of currency crises. *IMF Staff Papers*, (45): 1–48.

Kay, M., Franks, T. & Smith, L. (1997). *Water: Economics, Management and Demand*. London: E. & F.N. Spon.

Kemper, K. (1996). *The cost of free water: water resources allocation and use in the Curu valley, Ceará, Northeast Brazil* (Linköping, University of Linköping).
Kemper, K.E. (2001). Markets for tradable water rights. In: Meinzen-Dick, R.S. & Rosegrant, M.W. (2001a).
Kemper, K.E. & Olson, D. (2000). *Water pricing: the dynamics of institutional change in Mexico and Ceará*, In: Dinar, A. (ed) Brazil. pp: 339–357.
Kinnersley, D. (1994). *Coming clean: the politics of water and the environment* (London, Penguin).
Koubratova, M. (1997). Water pricing for irrigation during the transition to market economy in Bulgaria. In: Kay, M., Franks, T. & Smith, L. 1997. *Water: Economics, Management and Demand*. London: Spon Press.
Kraemer, A. (1995). Water resources taxes in Germany. In: Gale, R. & Barg, S. with Gillies, A. (eds) *Green budget reform: an international casebook of leading practices* (London, Earthscan) chapter 16.
Kuhn, T.S. (1970). *The Structure of Scientific Revolutions*, Chicago: University of Chicago Press.
Lant, C. (2003). 'Commentary'. *Water International*, 28(1): 113–115.
Lee, F. (1994). Full cost pricing, in Hodgson et al., *op.cit*. Vol. 1, 262–266.
Linn, J.F. (1983). *Cities in the Developing World.* New York: Oxford University Press.
Lott, J.E., Parkinson, R.J., Fuller, M.P. & Beattie, L. (1999). *Nitrates and Phosphates in Surface Water on Jersey: Sources and Land Management Strategies*, Plymouth: Department of Agriculture and Food Studies of the University of Plymouth.
Lundqvist, J. (1998). 'Avert looming hydrocide'. *Ambio*, 27(6): 428–433.
Lundqvist, J., Lohm, U. & Falkenmark, M. (1985). *Strategies for river basin management: environmental integration of land and water in a river basin*. Dordrecht, Reidel.
McDonald, A.T. & Kay, D. (1988). *Water resources: issues and strategies* (Harlow, Longman).
MacRae, D. & Whittington, D. (1988). Assessing preferences in cost-benefit analysis: reflections on rural water supply evaluation in Haiti. *Journal of Policy Analysis and Management*, 7(20): 246–263.
Maslow, A. (1954). *Motivation and Personality*. New York: Harper and Row.
Matthews, A. & Carter, A. (1999). An *Assessment of the Persistence and Mobility of Chlorthal Dimethyl and Chlorthal*. Keyworth: ADAS.
Meinzen-Dick, R. & Rosegrant, M.W. (1997). Water as an economic good: incentives, institutions, and infrastructure. In: Kay, M., Franks, T. & Smith, L. 1997. *Water: Economics, Management and Demand*. London: Spon Press.
Meinzen-Dick, R.S. & Rosegrant, M.W. (2001a). Overview. In: Meinzen-Dick, R. & Rosegrant, M.W. (eds) (2001b).
Meinzen-Dick, R.S. & Rosegrant, M.W. (eds) (2001b). *Overcoming Water Scarcity and Quality Constraints*. Washington, IFPRI.
Merrett, S. (1997). *Introduction to the economics of water resources: an international perspective*. London, UCL Press.
Merrett, S. (1999a). '*The regional water balance statement: a new tool for water resources planning*'. *Water International*, 24(3): 268–274.
Merrett, S. (1999b). The political economy of water abstraction charges. *Review of Political Economy*, 11(4): 431–442.
Merrett, S. (2001). Behavioural studies of the domestic demand for water services in Africa, submitted for publication to *Water Policy*.
Merrett, S. (2002a). Water for Agriculture: Irrigation Economics in International Perspective. London: Spon Press.
Merrett S. (2002b). Twelve theses on the cost and use of irrigation water. *Irrigation and Drainage*, 51(3): 265–268.
Merrett, S. (2003). 'Virtual water and Occam's Razor'. *Water International*, 28(1): 103–105.

Merrett, S. (2004). Farm-level drought management: an irrigation case-study from the UK. *Water Policy* (forthcoming).

Merrett, S. & Sharp, C. (1991). The denarius hypothesis: house price inflation in owner-occupied London. In: Satsangi, M. (ed) *Changing housing finance systems* (Glasgow, Centre for Housing Research, pp. 271–312).

Miloradov, M. & Marjanovic, P. (1998). *Guidelines for conducting water resources assessment*. Studies and reports in hydrology No. 55, Paris, UNESCO.

Morris, J. et al. (2003). The sustainability of irrigation in England and the impact of water pricing and regulation policy options. In: Maksimovic, C., Butler, D. & Memon, F.A. (eds) (2003). *Advances in Water Supply Management*. Lisse: Swets & Zeitlinger: 623–632.

Mu, X., Whittington, D. & Briscoe, J. (1990). Modelling village water demand behavior: a discrete choice approach. *Water Resources Research*, 26: 521–529.

National Rivers Authority (1993). *NRA water resources strategy* (Bristol, National Rivers Authority).

National Rivers Authority (1995). *1995–96 annual abstraction charges* (Bristol, National Rivers Authority).

National Rivers Authority Thames Region (1994). *Future water resources in the Thames region: a strategy for sustainable management*. Reading, NRATR.

Ofwat (1994). *1993–94 Report on the Cost of Water Delivered and Sewage Collected*, Birmingham: Ofwat.

Owen, D. (1994). *A Review of Techniques for Systematic Client Consultation*. World Bank: Southern Africa Department.

Oxford University Press (1979). *The Compact Edition of the Oxford English Dictionary Volume I*. Oxford: Oxford University Press.

Pearce, F. (2003). 'Replumbing the planet'. *New Scientist*, 7 June: 30–34.

Perry, C.J. (1996a). *The IIMI Water Balance Framework: a Model for Project Level Analysis*. Colombo: International Irrigation Management Institute.

Perry, C.J. (1996b). *Alternative Approaches to Cost Sharing for Water Service to Agriculture in Egypt*. Colombo: International Irrigation Management Institute.

Pigou, A.C. (1932). *The economics of welfare* (London, Macmillan).

Pigram, J.J., Delforce, R.J., Coelli, M.L., Norris, V., Antony, G., Anderson, R.L. & Musgrave, W.F. (1992). *Transferable water entitlements in Australia* (Armidale, Centre for Water Policy Research).

Public Services Department (2000). Water Made Clear: Guidance Notes on the Provisions of the Water Pollution (Jersey) Law, 2000. Bellozanne: PSD.

Rees, J. & Williams, S. (1993). *Water for life: strategies for sustainable water resource management* (London, Council for the Protection of Rural England).

Rees, J., Williams, S., Atkins, J., Hammond, C. & Trotter, S. (1993). *Economics of water resource management* (Hull, University of Hull, R&D note 128).

Reisner, M. (1993). *Cadillac Desert: the American West and its Disappearing Water*. New York: Penguin Books.

Ricardo, D. (1821). *On the principles of political economy and taxation* (Cambridge, Cambridge University Press, edited by P. Sraffa, 1951).

Robins, N.S. (2000). The Water Resources of Jersey: an Overview. Wallingford: British Geological Survey.

Robins, N.S. & Smedley, P.L. (1998). *The Jersey Groundwater Study*. Keyworth: British Geological Survey.

Robinson, J. & Eatwell, J. (1974). *An introduction to modern economics* (London, McGraw-Hill, revised edition).

Rogers, P. (1991). What water managers and planners need to know about climate change and water resources management. *Proceedings of the first national conference on climate change and water resources management*, US Army Corps of Engineers.

Rogers, P. (1992). *Comprehensive water resources management: a concept paper.* Washington DC, World Bank.

Rosegrant, M.W. & Binswanger, H.P. (1994). Markets in tradable water rights: potential for efficiency gains in developing country water resource allocation. *World Development*, 22(11): 1613–1625.

Rosegrant, M.W. & Gazmuri Schleyer, R. (1996). Establishing tradable water rights: implementation of the Mexican water law. *Irrigation and Drainage Systems*: pp. 263–279.

Rosegrant, M.W. & Meinzen-Dick, R.S. (1996). Water resources in the Asia-Pacific region: managing scarcity. *Asian-Pacific Economic Literature*, 10(2): 32–53.

Rosegrant, M.W. & Ringler, C. (1998). Impact on food security and rural development of transferring water out of agriculture. *Water Policy*, 1(6): 567–86.

Russell, B. (1996). *History of Western Philosophy and its Connection with Political and Social Circumstances from the Earliest Times to the Present Day*. London: Routledge.

Sandmo, A. (1987). Public goods. In: Eatwell, J., Milgate, M. & Newman, P. (eds) *The New Palgrave: a Dictionary of Economics*, pp. 1061–1067, London: Macmillan.

Schiffler, M. (1997). Intersectoral water markets: a solution for the water crisis in arid areas? In: Kay, M., Franks, T. & Smith, L. (1997).

Schiffler, M., Köppen, H., Lohmann, R., Schmidt, A., Wächter, A. & Widmann, C. (1994). *Water demand management in an arid country: the case of Jordan with special reference to industry* (Berlin, German Development Institute).

Seckler, D. et al. (2000). *World Water Supply and Demand: 1995–2025*. Colombo: International Water Management Institute.

Shiklomanov, I.A. (1997). *Assessment of water resources and water availability in the world*. Stockholm, World Meteorological Organization.

Shiklomanov, I.A. (2000). Appraisal and assessment of world water resources. *Water International*, 25(1): 11–32.

Simon, H.A. (1959). Theories of decision-making in economics and behavioural sciences. *American Economic Review*, 49: 253–283.

Simon, H.A. (1982). *Models of Bounded Rationality: Volume 1, Economic Analysis and Public Policy; Volume 2, Behavioural Economics and Business Organization*. Cambridge, Mass.: MIT.

Söderbaum, P. (1994). Environmental policy. In: Hodgson et al., Vol. I, pp. 193–199.

Steppacher, R. (1994). William Kapp, K. In: Hodgson et al., Vol. I, pp. 435–441.

Talbot, R. & Whiteman, R. (1996). *East Anglia and the Fens*. London: Cassell.

Tuddenham, M. (1995). The system of water charges in France. In: Gale, R. & Barg, S. with Gillies, A. (eds) *Green budget reform: an international casebook of leading practices* (London, Earthscan, chapter 14).

Turner, R.K. & Dubourg, W.R. (1993). *Water resource scarcity: an economic perspective* (Norwich, Centre for Social and Economic Research on the Global Environment).

Veblen, T. (1919). *The Place of Science in Modern Civilisation*. New Brunswick: Transaction Books.

Watson, R.T. et al. (2001). *Climate Change 2001: Synthesis Report of the Intergovernmental Panel on Climate Change*. Cambridge and New York: www.ipcc.ch/pub/syreng.htm

Whittington, D. (1998). Administering contingent valuation surveys in developing countries. *World Development*, 26(1): 21–30.

Whittington, D., Briscoe, J., Mu, X. & Barron, W. (1990A). Estimating the willingness to pay for water services in developing countries: a case study of the use of contingent valuation surveys in Southern Haiti. *Economic Development and Cultural Change*, 38(2): 293–311.

Whittington, D. & Choe, K. (1992). *Economic Benefits Available from the Provision of Improved Potable Water Supplies*. Washington DC: US Agency for International Development.

Whittington, D., Choe, K. & Lauria, D. (1997). The effect of giving respondents 'time to think' on tests of scope: an experiment in Calamba, Philippines. In: Kopp, R.J., Pommerehne, W.W. & Schwarz, N. (eds) *Determining the Value of Non-marketed Goods: Economic, Psychological and Policy Relevant Aspects of Contingent Valuation Methods*, Rotterdam: Kluwer.

Whittington, D., Davis, J. & McClelland, E. (1998). Implementing a demand-driven approach to community water supply planning: a case study of Lugazi, Kenya. *Water International*, 23(3): 134–145.

Whittington, D., Lauria, D.T., Choe, K., Hughes, J.A., Swarna, V. & Wright, A.M. (1993). Household sanitation in Kumasi, Ghana: a description of current practices, attitudes, and perceptions. *World Development*, 21: 733–748.

Whittington, D., Lauria, D. & Mu, X. (1991). A study of water vending and willingness to pay for water in Onitsha, Nigeria. *World Development*, 19: 179–198.

Whittington, D., Lauria, D., Okun, D.A. & Mu, X. (1989). Water vending activities in developing countries: a case study of Ukunda, Kenya. *Water Resources Development*, 5: 158–167.

Whittington, D., Lauria, D., Wright, A.M., Choe, K., Hughes, J.A. & Swarna, V. (1993). Household demand for improved sanitation services in Kumasi, Ghana: a contingent valuation study. *Water Resources Research*, 29(6): 1539–1560.

Whittington, D., Mu, X. & Roche, R. (1990). Calculating the value of time spent collecting water: some estimates for Ukunda, Kenya. *World Development*, 18: 269–280.

Whittington, D., Okorafor, A., Okore, A. & McPhail, A. (1990). Strategy for cost recovery in the rural water sector: a case study of Nsukka District, Anambra State, Nigeria. *Water Resources Research*, 26: 1899–1913.

Whittington, D., Smith, V.K., Okorafor, A., Okore, A., Liu, J.L. & McPhail, A. (1992). Giving correspondents time to think in contingent valuation studies: a developing country application. *Journal of Environmental Economics and Management*, 22: 205–225.

Whittington, D. & Swarna, V. (1994). *The Economic Benefits of Potable Water Supply Projects to Households in Developing Countries*. Manila: Asian Development Bank.

Willows, R.I. & Connell, R.K. (eds) (2003). *Climate Adaptation: Risk, Uncertainty and Decision-making*. Oxford: UK Climate Impacts Programme.

Winpenny, J. (1994). *Managing water as an economic resource* (London, Routledge).

World Bank (1975). *Issues in Village Water Supply*. Washington: World Bank.

World Bank Water Demand Research Team (1993). The demand for water in rural areas: determinants and policy implications. *The World Bank Research Observer*, 8(1): 47–70.

Zaman, M. (1996). Sustainable irrigation management through cost recovery system. In: Howsam & Carter, op. cit. pp. 221–227.

Websites

http://www.anglianwater.co.uk/
http://www.environment-agency.gov.uk/

Index

Printed in the United Kingdom
by Lightning Source UK Ltd.
103626UKS00001B/76-87

9 781843 390817